AF789301

Advances in Soil Science

Advances in Soil Science

B.A. Stewart, Editor

Editorial Board

J.P. Quirk
U. Schwertmann
G.W. Thomas
P.B. Tinker
R.J. Wagenet
B. Yaron

Advances in Soil Science

Volume 4

Edited by B.A. Stewart

With Contributions by
L.R. Ahuja, W.H. Gardner, D.R. Keeney,
K.L. Sahrawat, and I. Szabolcs

With 42 Illustrations

Springer-Verlag
New York Berlin Heidelberg Tokyo

B.A. Stewart
USDA Conservation & Production Research Laboratory
Bushland, Texas 79012, U.S.A.

Copyright is not claimed for works by U.S. Government Employees: "Characterization and Modeling of Chemical Transfer to Runoff" by L.R. Ahuja

ISSN 0176-9340

© 1986 by Springer-Verlag New York Inc.
Softcover reprint of the hardcover 1st edition 1986

All rights reserved. No part of this book may be translated or reproduced in any form without written permission from Springer-Verlag, 175 Fifth Avenue, New York, New York 10010, U.S.A.
The use of general descriptive names, trade names, trademarks, etc. in this publication, even if the former are not especially identified, is not to be taken as a sign that such names, as understood by the Trade Marks and Merchandise Marks Act, may accordingly be used freely by anyone.

Typeset by Ampersand Publisher Services, Inc., Rutland, Vermont.

9 8 7 6 5 4 3 2 1

ISBN-13: 978-1-4613-8614-8 e-ISBN-13: 978-1-4613-8612-4
DOI: 10.1007/978-1-4613-8612-4

Preface

The world needs for food and fiber continue to increase. Population growth in the developing countries peaked at 2.4 percent a year in 1965, and has fallen to about 2.1 percent. However, in many developing countries almost half the people are under 15 years of age, poised to enter their productive and reproductive years. The challenges to produce enough food for this growing population will remain great. Even more challenging is growing the food in the areas of greatest need. Presently the world has great surpluses of food and fiber in some areas while there are devastating deficiencies in other areas. Economic conditions and the lack of suitable infrastructure for distribution all too often limit the alleviation of hunger even when there are adequate supplies, sometimes even within the country itself.

World hunger can only be solved in the long run by increasing crop production in the areas where the population is growing most rapidly. This will require increased efforts of both the developed and developing countries. Much of the technology that is so successful for crop production in the developed countries cannot be utilized directly in the developing countries. Many of the principles, however, can and must be adapted to the conditions, both physical and economic, of the developing countries.

This series, *Advances in Soil Science*, was established to provide a forum for leading scientists to analyze and summarize the available scientific information on a subject, assessing its importance and identifying additional research needs. But most importantly, the contributors were asked to develop and identify principles that have practical applications to both developing and developed agricultures. It is not the purpose of the series to report new research findings because there are many excellent scientific journals for that need. Communications in scientific journals, however, are generally restricted to short and technical presentations. Therefore, *Advances in Soil Science* fills a gap between the scientific journals and the comprehensive reference books in which scientists can delve in depth on a particular subject relating to soil science. The positive response to the first volumes of the series shows that such a gap did indeed exist.

The ultimate aim of the series is to stimulate action: action to determine where there are arable soils, action to develop technology for more efficient crop production on these soils, action to reduce the risk of degrading these soil resources, and action to determine on which soils our research efforts should be concentrated. Without such action, the task of producing adequate food in the future may simply be too great.

There are many audiences to reach. While intended primarily for scientists and students of soil science, this series also provides technical information for anyone interested in our natural resources and man's influence on these resources. The reviews are written by leading scientists from many countries, and will therefore provide the reader with information from a wide array of conditions. Such information will be particularly useful to professionals working in areas with developing agricultures because the reviews summarize and assess the significance of the technical literature. While the primary purpose of the series is to present reviews, the first contribution in this volume is a history. Dr. Walter Gardner, a noted soil physicist and a past-president of the Soil Science Society of America, has written the history of soil physics. This interesting and informative contribution will serve soil science, and soil physics in particular, in a meaningful way. Perhaps it will also stimulate others to write histories of other areas of soil science so that future students and scientists of soil science can fully benefit from the knowledge and inspiration of earlier scientists.

B.A. Stewart

Contents

Contributors

L.R. Ahuja, Water Quality and Watershed Research Laboratory, Durant, Oklahoma 74702, U.S.A.

W.H. Gardner, Department of Agronomy and Soils, Washington State University, Pullman, Washington 99164-6420, U.S.A.

D.R. Keeney, Department of Soil Science, University of Wisconsin-Madison, Madison, Wisconsin 53706, U.S.A.

K.L. Sahrawat, International Crops Research Institute for the Semi-Arid Tropics, Patancheru P.O., Andhra Pradesh 502324, India

I. Szabolcs, Research Institute of Soil Science and Agricultural Chemistry, Hungarian Academy of Sciences, H-1525 Budapest, Hungary

Early Soil Physics into the Mid-20th Century

W. H. Gardner*

I. Introduction

The history of soil physics in its earliest manifestations is the history of soil science inasmuch as the earliest scientific observations and measurememts appear generally to be of soil physical properties. In fact, physical properties of soil have been noted in numerous cultural contexts, and it would be difficult to identify at just what point in historical time observations of such properties were sufficiently analytical, or measured with sufficient care, to be referred to as scientific. Hence, the best that can be said is that soil physics began in antiquity. Where and in what early culture it began would be equally difficult to determine and most histories inevitably will bear the mark of the cultural background of the writer. Moreover, the history of soil physics involves innumerable people, of diverse interests and backgrounds, increasing almost unmanageably by the beginning of the 20th century. By mid-century, the beginning of the contemporary period of the author's

*Professor Emeritus, Department of Agronomy and Soils, Washington State University, Pullman, Washington 99164-6420, U.S.A.

© 1986 by Springer-Verlag New York, Inc.
Advances in Soil Science, Volume 4

professional career, even the subfields of soil physics have become too extensive to cover adequately in a short history. Hence, selection of persons and topics to include becomes increasingly difficult and little attempt is made here at complete and comprehensive coverage, leaving the task to future historians with perspective improved with time. What is covered here must of necessity bear the mark of the writer and his own interests.

Background information about people in soil physics history is difficult to obtain, but of considerable interest to many readers. The writer views such information on outstanding contributors to the science to be worthy of publication wherever it is available. Brief notes along this line are given when available from the author's own experience, related by others, or available from literature.

That the early histories of soil physics and soil science are essentially the same derives from the fact that physical principles are involved somehow in most phenomena concerning soil. Furthermore, the early scientists, or natural philosophers as they are better known, were not specialized, specialization being a relatively modern historical phenomenon. What today would be identified as scientific endeavor often was an avocation of men largely engaged in other affairs, such as the law or as physicians. Much later in the unfolding of soil science compartmentalization developed, even to the extent that a few subfields have moved away and lost their identity as a major part of soils studies. Over the years interests have diverged and converged and diverged again as the tools of science have been brought to bear upon theoretical and practical problems perceived as important. The physicist, particularly, has found his subject relevant in some way to most of what constitutes modern soil science and research emphasis has followed various interests, with few central themes holding sway.

II. Early Beginnings

That primordial attention has been given to soil arises from the fact that soil is the porous body upon which civilizations are built. These civilizations have derived their support and sustenance from plants grown on soil and have depended upon water stored in soil and moderated in its flow to the sea by soil. Soil is formed from physical and chemical weathering of rocks—processes described historically because they involve eons of time—by glaciation, and by wind and water transport of soil materials, later deposited in deltas and loessial planes. There, soil undergoes further transformations over time and provides a habitat for biological life and a base for the development of civilizations.

Possibly the earliest attention to soil occurred before the dawn of

Figure 1. S. N. Kramer's (1959). *History Begins at Sumer*, p. 68.

written history, as may be inferred from Sumerian cuneiform writings from the cradle of civilization in the valley of the Euphrates River. In these writings on clay tablets, dating about 1700 BC, are found instructions on land preparation and planting of grain. One of these writings, reported by Samuel Noah Kramer (1958), begins, "In days of yore a farmer gave (these) instructions to his son: When you are about to cultivate your field, take care to open the irrigation works (so that) their water does not rise too high in it (the field). When you have emptied it of water, watch the field's wet ground that it stays even; let no wandering ox trample it. Chase the prowlers and have it treated as settled land. Clear it with ten narrow axes (weighing no more than) 2/3 of a pound each. Its stubble (?) should be torn up by hand and tied in bundles; its narrow holes shall be gone over with a drag; and the four sides of the field shall be fenced about." Writings on the tablets further describe breaking the ground up with implements that would be described as a mattock and hoe and pulverizing clods by "hammering." Furrowing and planting with a plow evidently was to be carried out with "straight" furrows, alternating annually (?) with "diagonal" furrows. A seeding "funnel" was attached to a plow pulled by oxen with seed being made to "fall two fingers uniformly." If the seed failed to penetrate the earth properly the "tongue of the plow" was to be changed. The plow and seeding implement, resembling in principle a modern planter, were clearly shown in a picture on one of the tablets. Vegetable gardens and fruit groves also are mentioned in the writings. A later and more complete translation of this material is provided by the same author (S.N. Kramer, 1963).

Some of the earliest historical references to soil 3000 or more years ago (Bennett, 1939) have to do with erosional forces of wind and water and their relevance to cultural activities. Physical properties of soil are intimately involved in biological processes, both affecting them and being affected by them, as has been noted by Sir E. John Russell (1957). Russell also describes an ancient practice, further developed and used in the 18th and 19th centuries, wherein fodder plants are grown on near-barren light-textured soil and then grazed by sheep. Their droppings manured the land and trampling of feet consolidated it. The benefits were so great that "farmers regularly spoke of the 'golden hoof' as the best amendment for light soils."

Numerous references to soil are found in historical writings. Buol *et al.* (1973) refer to writings involving soil by Aristotle (384–322 BC), Theophrastus (372–286 BS), Cato the Elder (234–149 BC), and by Varro (116–27 BC). In about 360 BC Democritus (460–370 BC) suggested that plant growth involved a cycle of indestructible elements, in keeping with his postulate that the universe consists of empty space and an, almost, infinite number of indivisible and invisible particles which differ from each other only in form, position, and arrangement. Later, Aristotle taught that plants absorbed through the roots from humus the necessary materials for growth (Salmon and Hanson, 1964). Sir E. John Russell (1957) quotes from Pliny Secundus (23 or 24–79 AD) on the use of marl (or chalk referred to in other writings, calcium carbonate) for nourishing the land. Charles E. Kellogg (1938) in the USDA Yearbook of Agriculture *Soils and Men*, refers to the Bible, Book of I Samuel, 13:20, which directs the Israelites "to sharpen every man his share, and his coulter, and his axe and his mattock"; and he refers also to a passage from Homer (800 BC or earlier) concerning Oddysseus, who, "upon returning from his wanderings, was recognized by his dog lying on a heap of refuse 'with which the thralls were wont to manure the land.'" He also refers to Varro, Pliny, Cato and, especially, Columella's treatese (about 60–65 AD)* of whom he says, "Although Columella remained the authority on agriculture for more than 14 centuries, he made no claim to originality, in the sense of invention." Kellogg also says that the Chinese made a schematic soil map of their country about 42 centuries ago as a basis for taxation and for the administration of agricultural affairs.

Cyril G. Hopkins (1910) refers to the "Georgical Essays, Edition of 1777, by Doctor A. Hunter," which cite some of the ancient writings referred to above but include: "Hesiod wrote very early upon agriculture" and "Mago, the Carthaginian general, composed 28 books on the same subject." These books, captured at the fall of Carthage to the Romans in 146 BC, were translated and became the most authoritative books on agriculture in early Rome.

---

*Kellogg's citation reads Columella, Lucius Iunius Moderatus. 1745. *De Re Rustica*. (Anonymous translation . . .) *Columella's Husbandry*, 600 pp., illus. London.

III. The Renaissance to the 18th Century

Relatively little is found in the literature between the first century AD and about the 1200s, when interest in the soil was renewed. Peter Crescentius, in Rome, published a book on agriculture in 1240. Bernard A. Keen (1931), in his book *The Physical Properties of the Soil*, provides an extensive bibliography of contributions to soil physics. He refers to *Fitzherbert's "Bok of Husbandry"* (1523) which he says is the "earliest work in the English language dealing expressly with practical agriculture" and contains many observations on soil cultivation.

One of the most remarkable of early natural philosophers was the Frenchman Bernard Palissy (1509?–1589), who evidently understood that vegetation derived from "salts" and water obtained from soil. His first reference to fertilizers was in a book referred to generally by the abbreviated title *Recepte Veritable* (1563). This book is "the description of a refuge for the just—that is, the persecuted Protestants—where they would live in peace as farmers and shepherds." In this he gives his opinion of the best methods of farming (Aurele LaRocque, 1957). Here and in his 1580 book, *Discours Admirables* (LaRocque, 1957), he recommends collecting and using the water from manures and describes how, when farmers pile manures on the land for later spreading, growth is better in those places where rainfall has carried materials from the manure piles into the soil. He devotes several chapters to discussion of "salts," the earth, and agricultural use of and exploration for marl (his usage: soil consisting of clay and carbonate of lime). Of "salts" he says, "I tell you that there are so many kinds that it is impossible for any man to name them all, and I tell you, moreover, that there is nothing in this world that does not contain salt, either man, beast, trees, plants or other vegetative things: even metals: and I say further, that no vegetative thing could vegetate without the action of salt, which is in seeds; what is more, if the salt were removed from the body of man, he would fall into dust in less than a wink."

In his explorations of rocks and soil he describes the construction and use of a "soil auger" with detachable handles of various lengths. And, in the dedication of *Discours Admirables* he says that certain books written in "beautiful Latin, or other well polished language have left many pernicious talents to delude youth and waste its time" and says that "Such pernicious books have led me to scratch the earth, during forty years, and to search its bowels, in order to know the things it produces within itself." He comments on use of trees and grass for erosion control, and of clay for lining ponds and streams to reduce water loss.

Palissy was trained in the manufacture of stained glass windows and was the inventor of enameled pottery, which led him to an extensive study of ceramic arts and particularly of the properties of clays and substances which modify its ceramic properties. He became a self-made paleontologist and a keen observer of nature. He seemed to be defensive of his lack of formal education but challenges the "opinion of so many famous and ancient philosophers, who have written on natural things, and filled the whole earth with wisdom." He says "I have set up a cabinet in which I have placed many and strange things which I have drawn from the bowels of the earth, and which give reliable evidence of what I say." Copies of the labels in his extensive collection are included in his book. He appears, possibly, to be the earliest scientific experimentalist.

Palissy was very much ahead of his time in theorizing that the origin of ground water and springs was rainfall which infiltrated the soil, obviously the soil playing an important role in conserving the water. In one place he says, "This must make you consider well that rain water that falls on mountains, lands, and all places that slope toward rivers or fountains, do not get to them so very quickly. For if it were so, all fountains would go dry in summer. From this it follows that under these rivers there are many continual springs, and in this way, not being able to flow except little by little, all springs are fed from the end of one winter to the next." The prevailing view, held by Plato (428–348 BC) and by Aristotle, by important natural philosophers of Palissy's day, and by a few even into the 19th century, was that water moved from the oceans to regions beneath the land surface, where it somehow moved upward into springs and rivers. Of this Palissy says, "If the rivers and fountains of the mountains proceeded from the sea, as some say, it would be necessary for the waters to flow from the sea in some place where it is higher than all the mountains, and that there should be a well enclosed pipe from the high seas to the summit of the mountains." Another popular theory of the day was that water was vaporized and moved upwards in caverns, where it condensed, thus supplying the underground water in wells and water flowing out from springs (Meinzer, 1934)

About 66 years after Palissy's book, in 1629 van Helmont (1580–1644), discoverer of carbon dioxide, was to perform his experiment wherein he grew a willow tree in a tub of soil and concluded from measurements that the gain in weight resulted from transmutation of water into plant tissue. This seemed to confirm a belief expressed by Sir Francis Bacon (1561–1626) in 1627 that water was the source of plant nourishment (Daumas, 1958). This theme was followed in 1673 by John Evelyn (1620–1706), an early secretary of the Royal Society of London, who gave several lectures to the Society dealing with improvement of the earth "for vegetation and propagation of plants."

More detailed experiments with soil began late in the 17th century and early in the 18th, as exemplified by J. Houghton's experiment described in 1706 by John Mortimer (1656–1736) (Keen, 1931). In these, Houghton "dissolved" some clay in water and poured the "thick" into a separate basin until all was gone but the sand. From such experiments he was able to infer the relative proportions of fine and coarse materials in different soils and to observe the differences this made in soil physical properties. Keen credits Houghton with expansion of ideas about soil through observing relations between it and its moisture retention, together with his observations on soil particle shapes and the tortuous nature of passages between particles of some soils, which interferes with water penetration.

IV. The 18th to Early 19th Century

Jethro Tull (1674–1740) of Oxford, educated as a lawyer but becoming a gentleman farmer in 1699, promoted the art of cultivation in the 18th century with his book, *Horse-hoeing Husbandry* (1731), extensively enlarged in 1733, and in numerous other writings. In these he describes drilling seed into rows and cultivating with a horse hoe in between (Hopkins, 1910; Keen, 1931; Wayne D. Rasmussen, 1960; and numerous others). In contrast to the teachings of Aristotle, that plants derived their nourishment from humus, Tull believed that finely pulverized soil constituted the food or "pabulum" of plants, hence his strong belief in the importance of cultivation. He ascribed the "main action of dung on the soil to the crumbling effects of the ferments it contains and asserted that the same result was obtained much more efficiently by tillage alone" (Keen, 1931). The absurdity with which such views would be taken today may be somewhat mitigated by a statement from Tull's writings, quoted by Hopkins (1910) as follows: "As to the fineness of the pabulum of plants, it is not unlikely that roots may insume no grosser particles than those on which the colors of bodies depend; but to discover the greatness of those corpuscles, Sir Isaac Newton thinks, will require a microscope that with sufficient distinctness can represent objects five or six hundred times bigger than at a foot distance they appear to the naked eye." Considering the primitive state of the art in molecular physics and chemistry in the early 18th century the main absurdity apparently lies not so much in his "pabulum" theory, but in his assumption that cultivation could pulverize a soil to such a fine state.

Numerous natural philosophers in the early years of modern science, near the beginning of the 18th century, contributed to the early development of the agricultural sciences. Among these were Joseph Priestly (1733–1804), discoverer of oxygen; Antoine Laurent Lavoisier (1743–1794), who in 1778 started a farm and conducted field experiments important to agricultural science, as well as contributing in a major way to chemistry and general science through his work on the composition of water; and Jean Senebier (1742–1809), who discovered the basic facts of photosynthesis in about 1782 (E.J. Russell, 1912). Lavoisier's perfection of a quantitative balance was a major contribution to soil science as to all sciences. Priestley and Lord Cavendish (1731–1810) demonstrated that atmospheric nitrogen could be combined with oxygen by means of an electric arc, thus anticipating a method for producing nitric acid used a century later. Cavendish measured the comparative weight of gasses arising from decomposition of plant and animal substances.

The beginnings of modern hydrology appear to be in France, where Pierre Perrault (1611–1680), a French lawyer with geology and meteorology as avocations, followed by Edme Mariotte (1620–1684), measured precipitation in the Seine river basin and estimated the amount to be many times that of the flow of the Seine.

However, it should be noted that Leonardo da Vinci (1452–1519), even before de Palissy and Perrault, had left a note in his book "On the Waters of the Earth" (Edward MacCurdy, 1938), which was only published after his time. This note indicated that da Vinci clearly recognized the power of the sun to vaporize water and something of the movement of air masses. His note says,

> Where there is life there is heat, and where there is vital heat there is movement of vapour. This is proved because one sees that the heat of the element of fire always draws to itself the damp vapours, the thick mists and dense clouds, which are given off by the seas and other lakes and rivers and marshy valleys. And drawing these little by little up to the cold region, there the first part halts, because the warm and moist cannot exist with cold and dryness; and this first part having halted receives the other parts, and so all the parts joining together one to another form thick and dark clouds.
>
> And these are often swept away and carried by the winds from one region to another, until at last their density gives them such weight that they fall in thick rain; but if the heat of the sun is added to the powers of the elements of fire, the clouds are drawn up higher and come to more intense cold, and there become frozen and so produce hailstorms.
>
> So the same heat which holds up so great a weight of water as is seen to fall in rain from the clouds sucks it up from below from the roots of the mountains and draws it up and confines it among the mountain summits, and there the water finds crevices and so continuing, it isssues forth and creates rivers.

Edmond Halley (1656–1742), for whom "Halley's Comet" is named, considered evaporation of the Mediterranean area, comparing estimates of evaporation of the Mediterranean based on his evaporation rate measurements with the inflow from rivers. His calculations showed that evaporation was three times the inflow from rivers (Halley 1687, 1691; Philip, 1977). John Dalton (1766–1844) in an essay (Dalton, 1793) described and discussed evaporation from land and water surfaces. He emphasized the importance of heat, dry air, and decreased pressure of the atmosphere to surface evaporation. His experiments, anticipating modern evaporation pan measurements and lysimetery, indicated that evaporation from land and water in the temperate and frigid zones was not equal to the rain that fell, even in summer. This gave excellent support to the hypothesis of de Palissy, more than two centuries earlier, that rainfall was the source of surface and underground waters.

Contributions and involvement of basic scientists from such fields as chemistry, physics, and mathematics, begun in the early years of soil science, continue today, particularly in the field of soil physics. Moreover, as will be evident later, many soil physicists have been educated primarily as physicists and identify strongly with that field. It is of some interest to note that the author's father, Willard Gardner, often proclaimed that there was no such thing as "soil physics," but only physics applied to the description of physical properties and processes involving soil. Parallels of appreciable interest exist in the histories of both soil physics and physics, particularly the dearth of scientific progress from about the second century AD until the time of Roger Bacon (1214–1292) and da Vinci.

Numerous people who have made important contributions to physics also are important in the field of soil physics and their names occur frequently in discussion of principles fundamental to the understanding of soil physical properties and processes. In his classic book *Introduction to Modern Physics* (1934), F. K. Richtmyer presents a survey of the origin and development of modern physics in which he breaks the history into five time periods: His *first period* extends from the earliest times to about 1550 AD, which date marks roughly the beginning of the experimental method. Some familiar names in physics mentioned for this period include Thales of Miletus (624–547 BC), Pythagoras (580–500 BC), Philolaus (470–399 BC), Democritus and Aristotle (referred to earlier), Euclid (last half of 4th century BC) and Archimedes (287–212 BC).

Copernicus (1473–1543), Francis Bacon, and da Vinci (referred to earlier) fit better into Richtmyer's *second period*, extending from 1550 to 1800, which includes an increasing number of names such as Tycho (1546–1601), Galileo (1564–1642), Kepler (1571–1630), Newton (1642–1727), Bernoulli (1700–1782), Franklin (1706–1790), Black (1728–1799) and Cavendish.

The *third period*, 1800–1890, is the period of classical physics, when a great many physicists believed that all of the important laws of physics had been discovered, and includes such names of interest in the field of soil physics as Laplace (1749–1827), Count Rumford (Benjamin Thompson) (1753–1814), Sir Humphrey Davy (1778–1829), Ohm (1787–1854), Peltier (1785–1845), Faraday (1791–1867), Fourier (1768–e1830), Carnot (1796–1832), Joule (1818–1889), Stokes (1819–1903), Helmholtz (1821–1894), Kirchhoff (1824–1887), Lord Kelvin (William Thomson) (1824–1907), Maxwell (1831–1879), Josiah Willard Gibbs (1839–1903), Stefan (1835–1893), Boltzmann (1844–1906), and Planck (1858–1947)

The *fourth period* begins with the discovery of the photoelectric effect, the history of which begins with Hertz (1857–1894), and is followed by Roentgen's (1845–1923) discovery of X-rays in 1895, radioactivity by

Becqueral (1852–1908) in 1896, the electron by J. J. Thomson (1856–1940) in 1897 and Albert Einstein's (1879–1955) photoelectric equation in 1905. These discoveries completely shattered the early belief that all important physical principles had been discovered and it remained only to refine the constants. It is in the fourth period when the sciences of soils and soil physics really began to grow. Here and, particularly in the period to follow, soil physicists began to apply physical principles, the tools of physics, and analytical methods to studies of soil. Richtmyer ends his fourth and begins his *fifth period* in about 1925, suggesting that future historians may regard the new subject of wave mechanics as exerting as profound an effect upon physics as did the discoveries of Sir Isaac Newton earlier.

In the first sentence of his book on *The Physical Properties of the Soil*, referred to earlier, Keen (1931) says that "The preparation of the soil by cultivation implements to secure good conditions for seed germination and plant growth is the oldest branch of arable agriculture." He devotes considerable space to this subject and credits James Small of Scotland and Thomas Jefferson (1743–1826), Third President of the United States and author of the U.S. Declaration of Independence, with the new development of a "geometrical method of mouldboard design," referred to as the "mouldboard plow" (1790). Jefferson's part in this development, and numerous other contributions identifiable as soil physics in early American agriculture (Betts, 1944, 1953), led the author to designate Jefferson as "the new nation's first practicing soil physicist" in a bicentennial paper, "Historical Highlights in American Soil Physics, 1776–1976" (Walter H. Gardner, 1977). Work with crop rotations and soil amendments, his careful observations of seeding, emergence, and growth of plants, recorded in detail in his "Garden" and "Farm" books and in his *Meteorological Journal* (appendix to the Garden book) (Betts 1944, 1953); and his technical correspondence with numerous farmer contemporaries, including such notable figures as George Washington and James Madison, the First and Fourth United States Presidents, gave impetus to soils investigations in America and amply justify his position as a contributor to soil science.

Tilling and plowing to "pulverize" the soil began to receive increased attention, particularly in England and in continental Europe, in the 17th and 18th centuries. Systematic experiments were conducted to improve the pulverizing action and to measure and reduce the drafts of plows. Jefferson sought a dynamometer for measuring the draft of his plow, and, while he was president he corresponded with Robert Patterson and borrowed a book from him to refresh his memory of "Emerson's fluxions" (Newton's term for the "differential" of calculus). He had studied these at William and Mary and, evidently, he thought they might have some use in the design and operation of a plow. At the same time interest in fertility problems was on the increase and these often were associated with pulverization and to the surface area of the soil particles. This led quite naturally into particle size studies or to the "mechanical analyses" that were to significantly occupy the attention of soil physicists, as well as other soil specialists, for all of the 19th century and well into the 20th, when soil water began to take over as the dominant theme of soil physics.

The rapid growth of the natural sciences in the 17th and 18th centuries heightened interest in soil as a chemical substance in which plants grew, so that early in the 19th century there began to develop an appreciable literature dealing with soil analyses as related to plant growth and with soil humus and fertilizers to reclaim "worn out soil." Europeans were well ahead of Americans in this because of the dominance of frontier farming on this continent, where fertile new lands were so easily obtained. Jefferson, in a letter to George Washington in 1793, had discussed how rotations involving small grain and red and white clover could be used to restore the fertility of "exhausted" lands following Indian corn (*Zea mays*) and tobacco (*Nicotiana tobacum*). He also had recognized the value of manures, but he indicated that "we can buy an acre of new land cheaper than we can manure an old acre (Betts, 1953)."

The coming of steam locomotion to the farm in the mid-19th century revolutionized plowing and some of the power intensive farm operations. With ample power, deep plowing and "rotary" cultivation, which did more than to merely turn the soil, became possible and agriculturists began to debate the merits of such operations as subsoiling. Questions raised and arguments closely resemble those continuing today over such matters as how long the effect of a certain deep tillage operation will last. Revival of an ancient practice of draining lands was accelerated as steam power made trenching a more practical process. Deep versus shallow drainage and the influence of cultivation on the wetness (and warming) of plow-depth soil was evaluated. From generalizations attempted regarding both tillage and drainage it is evident that soil was regarded as a much simpler material in these early times than it is today. Recognition of the complexity and variability of soil and the need for subjective evaluation would have simplified many of the debates of those days—an assertion which is not without some merit even at the present time.

V. The Birth of Soil Physics

By the middle of the 19th Century a number of American scientists, influenced largely by such Europeans as Sir Humphry Davy in England, had begun to apply the rapidly advancing sciences of chemistry and physics to the soil. Davy was the son of a woodcarver and was apprenticed to a surgeon-apothecary in Penzance, Cornwall, England.

As a young man of 23, having acquired a reputation as a lecturer at the Royal Institute of London, he was appointed professor of chemistry. His research in agriculture resulted in publication of an 1813 book, *Elements of Agricultural Chemistry* Other Europeans having considerable influence included, Gustav Schubler (1787–1834), Baron Justus Von Liebig (1803–1873), Wilhelm Schumacher, Alphonse Theophile Schloesing (1856–1930) in Germany, and Jean-Baptiste Josepoh Dieudunne Boussingalt (1802–1887) in France. Sir E. J Russell (1872–1965), in his book *Soil Conditions and Plant Growth* (1912, revised in 1950), reports that Davy's "insistence on the importance of the physical properties of soils—their relationship to heat and to water—marks the beginning of soil physics, afterwards developed considerably by Gustav Schubler."

Hans Jenny, in his book *E. W. Hilgard and the Birth of Modern Soil Science* (1961), indicates that "Chaptal was highly impressed by the experiments of Sir Humphrey Davy who had 'ascertained the comparative energy with which different soils absorb the moisture of the atmosphere; proving that those which are the most fertile possess this power in the highest degree; and so uniform is this rule, that the fertility of soils may be estimated and classed by it alone.'" Jenny refers to an 1845 book, *Chymistry Applied to Agriculture* by Count M. Chaptal which treats physical properties of soil. Leonard D. Baver, in his book *Soil Physics* (1940), also credits Davy with being among the first to recognize the significance of the physical properties of soils in agriculture, followed by Schubler, whom he credits with the "first technical investigations in soil physics," and who deals with the "physical properties which influence the productivity of soils" (Schubler, 1830). Schubler is credited by Keen (1931) with initiating the "first systematic study of the physical properties of soil. E. W. Hilgard, in his book *Soils* (1906), speaking of Schubler says, "He is really the father of agricultural physics." E. J. Russell (1912), quoting from Schubler's book, indicates that Schubler "ascribes the crumbling of calcareous clay soils to the difference in the contraction of calcareous sand and the clay substance. But it is doubtless more directly connected with the floculation of the latter by lime."

Baver (1940) refers to investigations by Schumacher, as reported in his 1864 book *Die Physik*, who uses Shubler's original data to develop ideas on the movement of air and water and introduces the "concept of capillary and non-capillary porosity" and "capillary-saturation capacity of soils." Schumacher considered capillary capacity to be a "function of the size of particle which determined the number and size of the capillary proes." "The rapidity with which water moved through the soil was visualized as dependent upon the structure of the soil as it affected the amount of non-capillary pores." Schumacher stressed the importance of the immediate soil surface to the entrance of air and water. He also called attention to the effects of the presence of a heavy layer below a permeable surface on the flow of soil water.

Figure 2. Darcy, Henri, (1803–1858). From *Journal of Petroleum Technology*, p. 12, 1956. Petroleum Branch, American Institute of Mining, Metallurgical and Petroleum Engineers. 800 Fidelity Union Bldg. Dallas, TX. Copyright ©. This school photograph of Henry Darcy was made at the age of 18 and is believed to be the only one presently in existence. A great nephew of Henry Darcy, Colonel Darcy, sent this picture to E. G. Trostel explaining that all others were destroyed during World War II.

Contemporary with Schumacher, but apparently unknown to workers in soil science, Henri Darcy (1803–1858, Figure 2) in 1856 published his now famous paper, *Les fontaines publiques de la ville de Dijon* (English translation: *The public fountains of Dijon*), in which the water flux through sand filter beds was indicated to be proportional to the gradient of the hydraulic head. The relationship, known widely as Darcy's law, has become one of the foundation stones in water-flow practice. A feature article on the life of Darcy was published in the *Journal of Petroleum Technology* on the 100th anniversary of the 1856 publication (George Fancher, 1956). Darcy was Chief Engineer in the city of Dijon, where he designed the water supply system, and later was Engineer and Chief Director of the Service of Water and Pavements in Paris. The first reference to Darcy in soils literature seen by the author occurs in King's 1901 book *Physics of Agriculture*, but it undoubtedly had been referred to earlier in physics or engineering literature inasmuch as King says, "The law of flow here referred to has been designated 'Darcy's Law'."

Two other contributions by 19th century scientists of importance in water-flow theory today, but rarely noted in early soils literature, are those of G. H. L. Hagen (1797–1884) in 1839 and of J. L. Poiseuille (1797–1869) in 1840 (referred to by King in 1901), who independently derived, starting with Newton's law of viscosity, an equation for water flux in capillary tubes in terms of the tube radius, the pressure gradient, and the viscosity of the fluid. This equation, known in recent literature as the Hagen-Poiseuille equation and which can similarly be derived for application to flow through other cross sections, expresses the same relationship as the Darcy equation, but with a conductivity term given in terms of the measurable quantities of radius and viscosity, which adds meaning to the expression. Numerous modern papers have been written explaining or extending the Darcy equation. Two of the many such papers were an analytical derivation of the Darcy equation based upon Newton's basic laws of motion and viscosity and extending the law to unsaturated flow by Warren A. Hall (1956) and a contribution by S. Irmay (1956) who extended the law to unsteady unsatuated flow.

Keen refers to the "Poiseuille" equation in his 1931 book and Joel E. Fletcher (1949) used the equation, rewritten to express the pressure as a combination of gravity and capillarity, to show by capillary analogy the influence on infiltration of such factors as pore size, temperature, surface tension, viscosity, wetting angle, and path length.

A discussion of early work on water flow in a porous medium like soil would not be complete without mention of Sir G. G. Stokes, Henry Selby Hele-Shaw (1854–1941), and Sir Horace Lamb (1849–1934). Their work on the theory of liquid flow around and over solid surfaces of various forms, not often cited in modern literature, has contributed appreciably to science. Stokes mathematical analyses form the basis for much theoretical work (Stokes, 1898) and the relationship of resistance to flow around a sphere, known as Stokes law, is the basis for the practical equation used in the pipette and hydrometer methods for particle size analysis. Hele-Shaw's dye-tracing techniques helped greatly to confirm the laminar nature of flow near surfaces, as would be experienced in ordinary water flow through porous bodies (Hele-Shaw, 1898). Lamb's book *Hydrodynamics* (Lamb, 1959) first appeared in 1879 under the title "Treatise on the Mathematical Theory of the Motion of Fluids." A 6th edition appeared long after in 1959. This book has been the basis for much modern development on fluid flow.

J. B. Boussingalt in France in 1834 likely was the first to work with field plots rather than the laboratory and pot experiments hitherto the fashion (E. J. Russell, 1912). His work largely concerned soil fertility but not exclusively, as shown by measurements he made of temperature beneath snow, at the snow surface, 11.9 m above the ground (Robert Warington, 1900).

Liebig in 1859 was highly critical of many farming practices, particularly those of the early American colonist, "who depoils his farm without the least attempt at method in the process. When it ceases to yield him sufficiently abundant crops, he simply quits it, and with his seed and plants, fetches himself to a fresh farm; for there is plenty of good land to be had in America; and it would not be worth his while to work the same farm to absolute exhaustion" (Hopkins, 1910). Thus, Liebig is recognizing, critically, what Jefferson had written to George Washington in 1793 about the lack of value of "manuring land." In the same year Hopkins reports another United States President, Abraham Lincoln, as recognizing the problem of despoiled farm land, suggested the need for studies of "deeper plowing, analysis of the soils, experiments with manures and varieties of seeds." Recognizing that population would increase rapidly he says, "erelong the most valuable of all arts will be the art of deriving a comfortable subsistence from the smallest area of soil." The problem of "despoiled farm land," addressed by Liebig and others, is today an increasingly serious problem discussed at length in numerous publications. Of particular interest are an English book (Nye and Greenland, 1960), *The Soil Under Shifting Cultivation*, and Charles E. Kellogg's 1963 paper, *Shifting Cultivation*.

The establishment of the Rothamsted Experimental farm at Harpenden, Herts, England in 1843 by John Bennet Lawes (1814–1900), proprietor of the Rothamsted Estate, and Joseph Henry Gilbert (1817–1901), a plant scientist brought in by Lawes, was of particular importance to research in soil physics. Rothamsted leadership under Sir A. Daniel Hall (1864–1942), ending in 1902 when he was appointed by Lloyd George to the new Agricultural Board, and by Sir E. John Russell (1872–1965) who succeeded him, placed heavy emphasis on soil physics, with appointments such as that of Sir Bernard Keen, R. K. Schofield, H. L. Penman, Russell's son E. W. Russell, and others. The importance of soil physical properties and processes have played a large role in the development of soil physics, particularly during the first half of the 20th century.

Martin Ewald Wollny (1846–1901, Figure 3) probably was the best known, if not the earliest, soil scientist to be called a soil physicist. As editor of the periodic journal, *Forschungen auf dem Gebiete Agrikulturphyskik*, 1878–1898, Wollny published numerous original articles and abstracts on soil physics, plant physics and agricultural meteorology. In original articles in his journal he showed that physical properties of soil studied by Davy and Schubler play a fundamental part in soil fertility

Figure 3. Wollny, Martin Ewald, (1878–1901). Photo courtesy of Don Kirkham, Iowa State University.

and he reported work on capillary rise (1884–85) with white chalk and soot to influence soil temperature (Woolny, 1878, 1881; E. J. Russell, 1912). Wollny introduced into the literature the term "crumb structure" (Krumelstruktur) for Hilgard's compound structure (Jenny, 1961). He made a strong case for production of crumb structure by earthworms and the building of "the humus surface soil layer," first suggested by Charles Darwin (1809–1882) (*The Formation of Vegetable Mold*, 1881, referred to by Hilgard, 1906). Wollny in his journal reports, "on the basis of observations of direct experimental cultures in boxes, with and without earthworms, surprising differences between the cultural results obtained." Production increases varied from 2.6 to 733% in his experiments (Hilgard, 1906)!

Among the American chemists who began serious work on soil analyses were Edmund Ruffin (1794–1865), recognized by some as the father of American soil chemistry, and Samuel W. Johnson (1830–1909), who played a major role in the establishment of the first American Agricultural Experiment Station at Connecticut in 1875 and was its director for 24 years. When Johnson was sent off to school at age 11 he encountered a world of books that completely absorbed his interest and led to his being permitted by his enthusiastic mother and tolerant father to build a "rude shack" near his father's farmhouse and to equip it as a chemical laboratory. This laboratory has been termed as the "nation's first experiment station," nearly 30 years before the establishment of the Connecticut station (Edward Jerome Dies, 1949). Johnson published a book, *Physical Properties of Soils as Affecting Soil Fertility*, in 1856 (referred to by Andrew Denny Rodgers, III, 1949), and in 1859 he reported his findings on uses of tillage, drainage, fertilizers, and the value of lime before the Smithstonian Institute in Washington (and in Experiment Station reports: Johnson, 1877, 1878).

Initially an advocate of chemical soil analyses, before his career was ended Johnson came to the view that only with extensive and careful work should it become possible to correlate soil chemical analyses with crop growth. However, he and a number of other American chemists of this era did initiate and popularize the application of scientific principles to the study of soil. In addition to his extensive work on chemical analyses, Johnson also considered physical properties of soils to have major importance and in an 1877 paper he discussed the purposes of tillage, citing increased absorption of rain, improved aeration, and reduction of evaporation and transpiration by weeds. In an 1878 paper he discussed capillary movement of water, anticipating principles that would later receive considerable attention: more rapid movement of water in moist pores than in dry, reduction of flow in fine pores, and the perceptive observation that "capillary properties of the soil must be evaluated in terms of rate as well as distance of movement," anticipating the role of dynamics in soil water retention clearly recognized only decades later.

Johnson was a frequent contributor, beginning at age 17 with an article "On Fixing Ammonia," to the American journal *Cultivator*, started in 1834 by Jesse Buol (1778–1839), whose motto was "To improve the soil and the mind." This journal, which later was succeeded by the *Country Gentlemen*, was subscribed to by most affluent farmers and carried articles citing European work of such scientists as Liebig and articles by American scientists. America's technical journals dealing with soils were to come later.

That clay usually is an important component of soil was known in antiquity; and many of its properties, such as those associated with ceramics, have been equally well known. However, its properties as a colloidal substance, particularly those involved with problems of water and plant growth, were less well known until late in the 19th century, when scientists such as Alphonse Theophile Schloesing (Leroux, 1931) and Jakob Maarten van Bemmelen (1830–1911) became interested. According to Hilgard (1906), Schloesing, in 1870 communications with the French Academy of Sciences, was the first to employ the term "colloidal clay." Van Bemmelen was able to show that soil had colloidal properties, including the property of retaining large amounts of water even when air dry (van Bemmelen 1878, 1879, 1910). Warington (1900), E. J. Russell (1912), and Rothamsted's Director Sir Daniel Hall [in an introduction to a 1925 lecture given by B. A. Keen (1926), "The Physicist in Agriculture"] all refer to van Bemmelen's early recognition of the colloidal properties of clay.

During the period of discovery of soil colloidal properties the role of microbial activity in soil also was demonstrated. Schloesing and A. Muntz, during a study of the purification of sewage water using soil filters, observed that ammonia in the sewage remained unchanged for about 20 days after percolation through columns had begun, after which time it began to be converted into nitrate. Finally, only nitrate passed through. Originally thought to be simple chemistry, it became evident that microbial activity was involved when it was observed that the process stopped with the addition of chloroform (E. J. Russell, 1912). Robert Warington (1838–1907); a professor at Oxford whose important 1900 book is cited earlier, followed these studies and began to apply similar principles to agricultural soils. Despite the attention he gives to soil physics Warington appears to be noted more for his nitrate studies than as a practicing soil physicist.

Schloesing studied colloidal clays intensively (1872, 1874) and, among other things, in 1872 he reported a discovery that quicklime precipitated clay. Eugene Woldemar Hilgard (1833–1916, Figure 4) in the United States and Eduard A. Mayer (1843–1942) in The Netherlands, in 1879,

Figure 4. Hilgard, Eugene Woldemar (1833–1916). From Vol. 1, *Soil Science*, 1916.

had independently made similar observations (Jenny, 1961). Since this early period studies of clay have been subdivided several times, from the points of view of both physical and chemical properties, and including, especially, their influence on practical problems of fertility and soil tilth.

VI. The Rise of Soil Physics, First Decade

Hygroscopic moisture became of considerable interest beginning in the latter part of the 19th century. In their 1912 publication *The wilting coefficient for different plants and its indirect determination*, Lyman J. Briggs and Homer L. Shantz refer to the 1859 observations of Julius von Sachs (1832–1897) showing a wide range of moisture contents of different soils at the time of wilting. Warington refers to Sachs (by name only) having taught that plants are able to make use of "hygroscopic water" in soils. This may well have been the first reference to "hygroscopic water" in the soils literature. Briggs and Shantz also refer to R. Heinrich (1894), as does Warington (again by name only) as having made the same observations as does Sachs. Warington reports that

> Heinrich grew plants in very small boxes till fully developed, and then placed them under conditions of very little evaporation till they began to wilt; the soil in the box was then mixed, and the proportion of water it contained determined. A variety of soils were employed. A weighed quantity of each soil was also placed in a dry state in a saturated atmosphere till it ceased to gain weight, and the amount of hygroscopic water which the soil could absorb was thus determined. It was found in every experiment, that when the plants wilted the percentage of water in the

Figure 5. Mitscherlich, Eilhard Alfred, (1874–1956). From Vol. 82, *Soil Science*, 1956.

soil was still somewhat higher than that proper to hygroscopic water only.

J. G. Mosier (1862–1922), University of Illinois, and Axel Ferdinand Gustafson (1880–?), Cornell University, in their 1917 book *Soil Physics and Management*, reproduce a table from a paper by A. F. Dobeneck (1892) that shows the water content of quartz and humus at relative humidities ranging from 30 to 100%. Hilgard (1906) quotes from "Mayer's *Agriculturchemie* (Vol. 2, p 131, 1905)," indicating that Mayer believes that "the hygroscopic efficacy of soils must be definitely eliminated from among the useful properties," but then indicates that Mayer "concedes the cogency of the experiments made by Sachs, which proved that dry soil immersed in a (probably not even fully) saturated atmosphere is capable of supplying the requirements of normal vgetation, thus explaining the obvious beneficial effects on vegetation of summer fogs." In 1901 Eilhard Alfred Mitscherlich (1874–1956, Figure 5) and in 1902 Herman Rodewald (1856–1938) improve things somewhat by reasoning that the amount of water vapor absorbed by a soil is proportional to the total surface and then attempt to calculate the surface on the basis that water is present as a monomolecular layer. By making the same measurements with toluene, Mitscherlich thought it possible to distinguish between internal and external surfaces (Baver, 1940). He attributed supreme importance to the surface offered by soil particles in determining the productivity of soils. In the preface of his 1905 book, *Bodenkunds fur Land-und-Forstwirthe*, moreover, Mitscherlich writes "it is absolutely indifferent to our cultivated plants how the soil on which they grow was geologically built up; the thriving of plants will always depend on the present physical and chemical condition of the soil." (Atanasiu, 1956). At least here, he possibly is not considering that the present physical and chemical condition of the soil may well depend upon its geological heritage. In his book he "claims that all determinations of soil hygroscopicity thus far made are grossly incorrect on account of the dew liable to be condensed on the soil layer from fully saturated air, as the result of slight changes of temperature" (Hilgard, 1906). Mitscherlich then suggested making such measurements over a 10% solution of sulfuric acid.

Mitscherlich was descended from a long line of scientists. His father was a professor of surgery and his grandfather was the famous chemist, Eilhard Mitscherlich, of the University of Berlin. His dissertation, completed under Rodewald at Kiel in 1898, was on the heat of wetting of soils. His work along these lines (Mitscherlich, 1901) and that of Rodewald (1902), were important contributions to the understanding of hygroscopicity of soils that prevailed until about two decades later.

Mitscherlich, Schloesing, and Wollny were the dominant European figures in soil physics at the end of the 19th century. Hilgard in Mississippi and California, Thomas Burr Osborne (1859–1929) in Connecticut, F. Hyrum King (1848–1911, Figure 6) in Wisconsin, and Milton Whitney (1860–1927, Figure 7), with the U.S. Department of Agriculture in Washington, dominated United States soil physics. Osborne, the son in law of S. W. Johnson, was most noted for his work on amino acids and discovery of the first vitamin. However, his method for mechanical analyses was widely used and he was quoted extensively regarding this most important subject of the day. However, Osborne's contributions (1887) were somewhat eclipsed early in the 20th century by the work of Sven Oden (1888–1934) in 1915 and 1925 (Lundegardh, 1934), who published extensively on the same subject.

Hilgard was born in Bavaria but came to Illinois at the age of 3. His early education was obtained with his father at home, but he went to

Figure 6. King, Franklin Hyrum (1848–1911). Photo supplied by Champ Tanner at the University of Wisconsin.

Figure 7. Whitney, Milton, (1860–1927). From Proc. and Papers of the First International Congress of Soil Science, Wash. D.C., 1927.

Switzerland and Germany at age 16 to study geology and chemistry, returning in 1855 at age 22 to Mississippi with a Heidelberg Ph.D. as State Geologist and Professor of Chemistry and Agricultural Chemistry. Eighteen years later he went to the University of Michigan as Professor of Geology and Natural History and 2 years after that, in 1875, needing a warmer climate for his health, he went to California. In Berkeley he quickly organized the agricultural work of the University of California and founded the Agricultural Experiment Station in 1875 which he directed along with being Dean of the College of Agriculture until his retirement to Emeritus status in 1906. The journal *Hilgardia*, replacing an Experiment Station publication in 1926, carries his name. Hilgard's earliest scientific paper (1860) was a *Report on the Geology and Agriculture of the State of Mississippi*, the last half of which was devoted to soils and their nature and management. He had important experience with and ideas about soil survey which were ignored in the early Bureau of Soils' establishment of a national soil survey, but these were readily identifiable in many of the later developments of the survey. His major contribution likely was his work on the relations of soil to climate (Bulletin 3 of the Weather Bureau, USDA, 1892, was on this subject) and his extensive contributions to knowledge of arid-area agriculture. A Russian contemporary of Hilgard's, V. V. Dokuchaev (1846–1903), agreed that "the properties of a soil depend not only upon its parent material but also on the climatic, vegetation and other factors to which it has been subjected" (E. J. Russell, 1912).

Jenny, (1961) says that "Hilgard's contributions to the physics of soil moisture were not as extensive and not in the same high plane as his works on soil chemistry and pedology." However, he adds that Hilgard "passed up a valuable lode. Every time he conducted its determination [minimum water capacity] a fundamental truth was staring in his face: water will not readily move from a moist soil (at field capacity) to a dry soil." His 1906 book *Soils*, was a major contribution to the entire field of soils, but particularly to soil physics.

F. Hyrum King, after having taught high school; served on a geological survey; studied physics, chemistry, biology, and geology; assisted in an ornithological study; and served as a professor of natural science, in 1888 joined the faculty at the University of Wisconsin as professor of agricultural physics. In 1901 he left Wisconsin to work in the U.S. Bureau of Soils, but returned to Wisconsin in 1904. His work at Wisconsin was across the field of agriculture but his seven books and numerous papers probably contain more information in the field of soil physics than in any other area (Tanner, 1964). His 1895 book *The Soil*, his 1899 book *Irrigation and Drainage*, and his 1901 book *Physics of Agriculture* remained the most authoritative texts on soil physics for several decades. Much of what he says about the origin of soil and its physical properties, soil moisture, aeration, temperature, tillage, and movement of water (King, 1892, 1898) is qualitatively correct by present day standards. The most significant shortcoming is lack of any quantitative treatment of the energy state of soil water, although he does discuss energy associated with the capillary lift of water.

King was regarded highly by his contemporaries and in 1904 Hilgard wrote in *Science* (Hilgard, 1904): "Both American and European scientists have been accustomed for many years to regard with confidence and respect the work and publications of the man whom, by common consent, the mantle of Wollny has fallen."

King's interests in putting water flow on a more quantitative basis is illustrated by his having involved his colleague, Charles S. Slichter (1864–1946, Figure 8), Professor of Applied Mathematics at the University of Wisconsin, in a flow problem that led to Slichter's 1898 paper "Theoretical investigation of the motion of ground waters." In this paper "an attempt is made to derive from purely theoretical considerations an expression for the flow of water or other fluid through a column of sand made up of grains of nearly uniform size and of approximately spherical form." Slichter shows "that there exists in the case of ground water movement what is known as a potential function, from which we may derive, in any determinate problem, the velocity and direction of flow, and the pressure at every point of the soil or rock." He applies these principles to discussion of water flow in horizontal planes and problems of the flow in wells, where vertical flow also is involved. It is interesting to note that Slichter expresses his flow equations in cgs units and then partially converts to English units, leaving the diameter of soil particles in millimeters, thus showing that the problem of units is not particularly modern. This paper introduces the physics term "potential" into the soil physics literature and appears to be one of Slichter's only six publications

Figure 8. Slichter, Charles Sumner, (1864–1946). (From Mark H. Ingraham, 1972. © University of Wisconsin Press. Reproduced with permission.)

in the field of soil (and water) physics; however, it is cited frequently in the literature over the next half century and, occasionally, today.

Slichter's greatest fame was as a teacher, graduate dean, and the founding genius of the highly successful Wisconsin Alumni Research Foundation. He truly was a most colorful "institution" at Wisconsin as books about and by him attest [H. Mark Ingraham,*Charles Summer Slichter, the Golden Vector (1972); Science in a Tavern*, by Slichter himself (1966), which includes much about the "Royal Philosophers," a "more or less formal group of diners" formed in about 1645 and which held most of its meetings in London taverns, and which was chartered in 1662 as the "Royal Society of London," and which included as a long-time secretary, John Evelyn who wrote and lectured extensively on agriculture and, particularly, about soils]. Herbert F. Wang, in paying further tribute to Slichter as a hydrologist, has referred to him as "an engineer in mathematician's clothing" (1986).

Slichter's insight into fluid-flow phenomena is illustrated by his application of the Laplace equations to the flow problem addressed by Darcy 40 years earlier. He wrote that "we have, therefore, shown that a problem in the steady motion of groundwaters is mathematically analogous to a problem in the steady flow of heat or electricity, or to a problem in the steady motion of a perfect fluid." Some of the earliest work on water flow is reported in an 1883 paper, "An experimental investigation of the circumstances which determine whether the motion of the water shall be direct or sinuous, and of the law of resistance in parallel channels," by Osborne Reynolds (1842–1912) of "Reynold's number" fame.

Milton Whitney probably was the most influential of all United States soil scientists over a three-decade period because of his position as director of the prestigious U.S. Bureau of Soils. At the same time he likely was the most controversial. He was born and educated in Maryland, taking a degree from Johns Hopkins University. He served briefly at the Connecticut and North Carolina Experiment Stations and became physicist and vicedirector at the Maryland Agricultural Experiment Station in 1888. Here, with U.S. Weather Bureau funds, he completed investigations leading to a paper, *Some Physical Properties of Soils in Their Relation to Moisture and Crop Distribution* (Whitney, 1892), which so impressed the agricultural officials with soils problems that they set up a Division of Soils in the Weather Bureau in 1894 with Whitney at its head. This division in 1901 became the Bureau of Soils under Whitney. The early publications and reports of the USDA and the State Agricultural Experiment Stations constituted the most important part of the United States scientific literature of soils in these early days.

Whitney's interests ranged widely over the field of soils and his and the Bureau's official position on soil fertility, expressed in Bulletin 22 (Whitney and Cameron, 1903), is particularly interesting:

> It appears further that practically all soils contain sufficient plant food for good crop yield, that this supply will be indefinitely maintained, and that the actual yield of plants adapted to the soil depends mainly, under favorable climatic conditions, upon the cultural methods and suitable crop rotation, a conclusion strictly in accord with the experience of good farm practice in all countries, and that a chemical analysis of a soil, even by these extremely delicate and sensitive methods, will in itself give no indication of the fertility of this soil or of the probable yield of a crop, and it seems probable that this can only be determined, if at all, by physical methods, as it lies in the domain of soil physics.

King, a physicist on the USDA staff from November 1901 to June 1904, arrived at a contrary position as a result of a comprehensive study of soil fertility, some of the data of which were used in Bulletin 22. This work resulted in six publications, the first three published as Bulletin 26 (King, 1905) and the other three rejected and published privately by King (1904) after his forced resignation from the Bureau in 1904. Whitney's letter of transmittal to the Secretary of Agriculture for Bulletin 26 indicates that the "facts presented are interesting and suggestive, and will be helpful to students of the soil; and seem, therefore, to call for publication, notwithstanding the fact that the opinions and conclusions which have been drawn from these facts must be considered as the personal views of the author, and in the main do not carry the indorsement of this Bureau."

Numerous non-Federal soil scientists, including one at Rothamsted in England, were upset by Bulletin 22 and what they regarded as unwarranted authoritarianism by the Bureau. Hilgard, one of the foremost critics, published several articles in *Science* (Hilgard, 1903, 1904) decrying what was happening. However, fear of upsetting a pending bill in the U.S. Congress, which would appropriate badly needed money for basic research (the Adams Act, which has since had great importance in United States agricultural research), cooled off what might otherwise have been a serious challenge to the operation of the U.S. Bureau of Soils. Cyril Hopkins, in his 1910 book *Soil Fertility and Permanent Agriculture*, discusses these events in detail, presenting numerous documents on both sides of the issue, including some testimony given by F. K. Cameron in 1908 (Hopkins, 1910) before the U.S. House of Representatives Committee on Agriculture that espouses the same theory on the inexhaustible nutrient supply in soils. This affair of three-quarters of a century past illustrates some of the prevailing concerns of soil scientists in the early days of soil physics and is of interest even today. Prevailing views today about soil stand in sharp contrast to Whitney's (1909) optimism regarding the soil resource given in Bulletin 55. "The soil is the one indestructible, immutable asset that the nation possesses. It is the one resource that cannot be exhausted; that cannot be used up." This is followed by an explanation of how fertilizers and rotations "act on or change toxic conditions in the soil, rendering the soil again sweet and healthy for the growing crop." Although the large effects of both soil fertility and soil physical properties (including soil water) on plant growth are well documented today, the relationships between soil chemical and physical analyses and plant growth have not yet been definitively established because of the complexity of growth processes and the large number of variables—chemical, physical, biological, and meteorological—involved. Soil heterogeneity, too, was recognized in this era as an important factor in assessing influences on plant growth (J. Arthur Harris, 1915, 1920), a subject to be studied intensively from a physical point of view by soil physicists later in the 1970s and 1980s.

Much of the growth of "soil physics" over the 19th century was external, England and Germany contributing significantly to the advances. However, at least one Englishman, Robert Warington, Professor of Rural Economy at Oxford, in the introduction to his *"Lectures on Some of the Physical Properties of Soil* (1900), reports that "The reader will probably be surprised that so little is said respecting English soils and so much respecting the soils of America. The writer heartily wishes that this might have been otherwise. In fact, however, the physical constitution and properties of English soils have as yet not been investigated, save in a very few exceptional cases." Warington quotes extensively from such early American soil scientists as Hilgard, R. H. Loughridge (1843–1917) (Hilgard's associate at California and who reported some of the earliest work on capillary rise (1892–1894)), Whitney, King, and Briggs, as well as such Europeans as Wollny, Mayer, Schloesing, and Schubler.

Major parts of early interests in soil were related to growing crops so that numerous biologists worked directly or indirectly with soil. Water

Figure 9. Briggs, Lyman J. (1874–1963). From a photograph supplied by the National Bureau of Standards.

conditions in soil were of considerable interest to an early plant physiologist, Burton E. Livingston (1875–1948), who with L. A. Hawkins in 1915 published a paper on "The water-relation between plant and soil." Earlier, in 1905, Livingston, J. C. Britton, and F. R. Reid had studied and reported on properties of unproductive soils. Livingston was with the U.S. Bureau of Soils in the years 1905 and 1906.

The continuing influence of the Bureau of Soils on soil physics into the 20th century is evident in the work of Lyman J. Briggs (1874–1963, Figure 9), who joined the Department of Agriculture as a physicist under Whitney in 1896. Briggs, a native of Michigan, did not attend high school but entered Michigan State College by examination at the age of 15, earning his B.S. degree at 19. He obtained an M.S. degree at the University of Michigan 2 years later and completed his doctoral degree at Johns Hopkins in 1901 while employed in the Weather Bureau of the USDA. Briggs' first contribution to the literature of soil physics came in his first year at the Weather Bureau, when he published his famous Bulletin 10, *"The Mechanics of Soil Moisture"* (Briggs, 1897) that asserted to present "the application of certain dynamical principles to the problems attending the movement and retention of soil moisture." The paper followed prevailing ideas regarding the existence of three forms of water—gravity, capillary, and hygroscopic— and described how capillarity was responsible for retention of water in interstices between soil particles and how the amount of retention would vary with the vertical height of a soil column. It failed to stress the dynamical problem of water flow, as did two subsequent papers, *"The Moisture Equivalents of Soils"* [Briggs and John W. McLane (1868–?), 1907] and *"The Wilting Coefficient for Different Plants and Its Indirect Determination"* (Briggs and Shantz, 1912).

Homer L. Shantz (1876–1958) served with the U.S. Bureau of Plant Industry for about 12 years, where, as a botanist, he worked closely with Briggs on plant–water relations. He was noted for his studies of plants in Africa, spending much time there. Later, from 1928 to 1936, he served as president of the University of Arizona and apparently had a difficult time during the depression as he supported faculty salaries in a period when downward adjustments were common. His secretary, many years later, reported to the author that he was a better scientist than president—always wanting to escape to his science office.

Briggs' and McLane's 1907 paper described a method for removing water from soil samples by a fixed and reproducible force as supplied by a centrifuge exerting a force of 3000 times gravity (later changed to 1000 g). By this method it was then assumed to be "possible to determine the retentive power of different soils for moisture when acted upon by the same definite force, comparable in magnitude with the pulling force to which the soil moisture is subjected in the field." An empirical equation was developed from observations on 104 different soils to relate the mechanical composition, as determined from particle size measurements, to the moisture equivalent. The 1912 paper describes a method for the indirect determination of the wilting coefficient and includes empirical equations for relating the wilting coefficient to the moisture equivalent, the hygroscopic coefficient, the moisture-holding capacity (water retained against gravity in a 1-cm-high cylinder set in water to a depth of 1 mm) and the mechanical composition in terms of sand, silt, and clay. The "maximum available moisture" for plant growth was the difference between the moisture-holding capacity and the wilting coefficient, a number "far in excess of that found in drained soils under field conditions." Although not explicitly designated as such in the paper, the moisture equivalent came to be widely regarded as *field capacity* and the maximum available water was the difference in moisture equivalent and the wilting coefficient.

The simplicity of these ideas and the tremendous utility they seemed to have led to almost universal adoption, persisting even today (with 1/2 or 1/3 bar percentage replacing the moisture equivalent) among many users of soil–plant–water information. In the context of the times these ideas constituted an important step forward but, as it has turned out, their apparent simplicity and utility diverted attention from the true nature of the soil–water system. It was many years later before the dynamics of the system began to be recognized, notwithstanding the fact that Briggs' contemporary and associate at the United States Bureau of Soils, Edgar Buckingham (1867–1940; Figure 10), published a paper in 1907, *Studies on the Movement of Soil Moisture*, which contained the basic ideas of water flow on which modern water-flow theory is founded. Many are inclined to place the blame on Briggs and the endorsement of Briggsian ideas by

Figure 10. Buckingham, Edgar (1867–1940). Photo, likely made from a color slide, courtesy of his daughter, Katherine Buckingham Hunt.

the prestigious research organizations of the USDA for lack of early attention to soil water dynamics. However, the basis for further development of soil water dynamics was available and in a U.S. Bureau of Soils publication contemporary with those of Briggs (Buckingham, 1907)! It is far more likely that soil scientists generally were unaware of the need for anything more comprehensive than Briggs' empirical soil moisture constants and that Buckingham's ideas merely were ahead of their time.

It is interesting to note that Buckingham moved from the U.S. Bureau of Soils to the U.S. Bureau of Standards in 1905 and that, after heading the biophysical laboratory of the Bureau of Plant Industry until 1917, Briggs joined the Bureau of Standards staff on a temporary basis for war work, much later, in 1933, becoming its director. Briggs had a long and productive career in the Bureau of Standards, from which he retired as director in 1945, but not from productive work which continued almost until his death in 1963 at the age of 88. Among notable achievements was his service in the development of atomic energy. He was chairman of the first government committee concerned with the military value of atomic fission, appointed by President Roosevelt to this post in 1939. He was an expert on aeronautics and beginning in 1917 he served for many years on the National Advisory Committee for Aeronautics (the predecessor of NASA) and was its vice chairman in 1942. In 1950 he returned to his much earlier interests in soil physics, with laboratory work that led to publication of "Limiting negative pressure of water" (Briggs, 1950), which was followed by similar studies on other liquids. In a response given at a luncheon commemorating his 80th birthday he gave as his third rule for a long life, "arranging of your work so as to try hard to avoid, if possible, working under pressure. This is very important and in my case, I have succeeded in working under negative pressure."

From the literature Briggs and Buckingham appear to have collaborated only once in their long association, first in the U.S. Bureau of Soils and later at the U.S. Bureau of Standards, and that was during World War II, when Briggs was asked to evaluate the bouncing characteristics of a baseball with the objective of using a substitute material for critical rubber in the balls which were supplied to military training camps. Briggs made the analyses with the assistance of "Dr. Buckingham on the theoretical considerations" (Cochrane, 1966). The author's correspondence with the daughter of Dr. Buckingham, Katharine Buckingham Hunt, has revealed that Drs. Briggs and Buckingham were close personal friends over their many years' association. She reports also that her father disliked administrative work and that her "mother could always tell when Dr. Briggs was away and he had to take his place."

Edgar Buckingham's involvement with soil physics lasted officially but three years (1902–1905) and produced two papers in this period. The first was on flow of oxygen and carbon dioxide in soil (1904) and the second (1907) appeared after he had left the Bureau of Soils to join the staff of the Bureau of Standards. His work at the Bureau carried him in different directions, but he did return at least once to a study of fluid flow, which resulted in a paper, "On plastic flow through capillary tubes" (1921).

Buckingham was born in Philadelphia but received his secondary school education in Newton, Massachusetts. He earned an A.B. degree in physics at Harvard College in 1887, did graduate work there for 2 years, and was an assistant in physics in 1888–1889 and 1891–1892. He studied at the University of Strasbourg and University of Leipzig, obtaining his Ph.D. from the latter in 1893. He taught physics and physical chemistry at Bryn Mawr College (Pennsylvania) in 1893–1899 and was an instructor in physics at the University of Wisconsin in 1901–1902, after which he joined the Bureau of Soils staff. Despite the significant contribution of his 1907 soil physics paper, he probably is best known for his work in fluid mechanics and thermodynamics, which he taught for a short time at the U.S. Naval Academy. He was the author of a pioneering book on thermodynamics published in 1900 and contributed widely in many scientific fields, pioneering on steam turbines and propeller design and in use of dimensional analyses. His high stature with the Bureau of Standards was indicated by the award of "independent status" to continue his work on theoretical thermodynamics in 1923.

In the 1907 paper Buckingham defined a "capillary potential," which in modern terminology would be referred to more frequently as "matric potential" or "moisture potential," and indicated that flow of water in soil would be proportional to the capillary potential and the "capillary conductivity," which should depend largely on the water content of the soil. In view of their importance to modern treatment of water flow it is instructive to consider at first hand some of what Buckingham had to say about the capillary potential and capillary conductivity in his monumental 1907 bulletin. About capillary potential he says,

> We shall assume that if we could, by purely mechanical means, pull a definite mass of water away from a definite mass of moist soil of a given moisture content, we should have to do a definite amount of mechanical work; and that if we then let the water and the soil come together again in obedience to their mutual attraction, we should, in principle at least, and if we could construct appropriate mechanism, be able to get back the same amount of work that we had to do in separating the water from the soil. This amounts to assuming that the attractive forces between the soil and the water are conservative, or that they have a potential.

After some discussion he then designates the potential as ψ and says, "The value of ψ, for a given state of packing, temperature, etc., depends only on the water content, decreasing as ψ increases. When the soil is completely saturated with water, its pores being full, water will begin to drain away from it at the first opportunity; it takes only an infinitesimal amount of work to remove a finite mass of water or $\psi = 0$. After further discussion he says, "The capillary potential for a given water content varies from soil to soil; the retentiveness of different soils, or even of the same soil in different states of structure, is different."

Of capillary conductivity Buckingham says,

> Let Q be the capillary current density at any point—i.e., the mass of water which passes in one second through 1 sq.cm. of an imaginary surface perpendicular to the direction of flow. Let ψ be a quantity which measures the attraction of the soil at any given point for water. Then the gradient of attraction, which we may denote by S, is the amount by which ψ increases per centimeter in the direction of the current, by reason of the fact that the water content of the soil decreases in that direction. Let ψ denote the capillary conductivity of the soil. Then we may write, in formal analogy with Fourier's and Ohm's laws, $Q = \lambda S$.
>
> The analogy, however, is only formal. In the first place, the thermal and electrical conductivities of a given piece of material are independent of the strength of the current and, in general, only slightly dependent on the temperature and other outside circumstances, so that for most purposes they may be treated as constants. The capillary conductivity, however, we have every reason to expect to be largely dependent on the water content of the soil, and therefore variable, not only from point to point in the soil, but also with the time at any given point. For it is not to be expected that the ease with which water flows through the soil will be independent of the extent and thickness of the water films through which—i.e. along which—it has to flow.
>
> Furthermore, the other factor in the equation, namely, the gradient S, is not the space variation of a simple and directly measurably quantity like a head of water, an electrical potential, or a temperature. It is the gradient of a quantity ψ, the attraction of the soil for water; and ψ depends in some as yet unknown way, differing from soil to soil, on the water content of the soil,

which can itself be measured only by tedious and not very accurate methods.

L. A. Richards, in his 1960 *Advances in Soil Physics*, quotes the same material on capillary conductivity from Buckingham's bulletin.

Despite its relevance to dynamic problems of water flow in soil, Buckingham's 1907 paper largely was ignored for more than a decade. Then Willard Gardner (1883–1964) and his co-workers and students, following the lead of Buckingham and Slichter, began to treat water flow in soil on a quantitative basis using potential theory.

VII. The Rise of Soil Physics, Second Decade

The earliest successful mathematical model of infiltration appears to be that of W. Heber Green (1868–1932) and G. A. Ampt (1887–1953), who wrote equations for vertical upward and downward and for horizontal flow, based upon the Hagen-Poiseuille equation applied to a bundle of capillary tubes and under the assumption that the entire pore space was filled as water advances (W. H. Green and Ampt, 1911, 1912). They indicated that saturation might be realized only for vertically downward flow and showed a water distribution curve for horizontal flow illustrating the departure from this assumption. They mention that Darcy (1856), Allen Hazen (1869–1930) in 1890, King and Slichter (1899), Bell and Cameron (1906), and Leather (1908) had made measurements finding that, "with certain limitations both the Poiseuille and Meyer-Poiseuille formula hold good." They refer also to Briggs' and Buckingham's papers. In a subsequent paper they deal with permeability of an ideal soil to air and water (W. H. Green and Ampt, 1912). E. W. Washburn, in 1921, wrote equations for the advance of water into porous ceramic materials which included such factors as viscosity of the displaced air, the angle of contact, and the coefficient of slip between water and solid. When these functions are neglected the equations agree with those of Green an Ampt. Philip (1954a, 1957), Childs (1967), W. R. Gardner (1967), Swartzendruber (1974), and others usefully have considered the Green and Ampt approach in dealing with infiltration. A book on ground water by Konrad Keilhack (1912) was published in Germany during this same period.

Philip (1983) discusses the limitations of the Green and Ampt equations in light of modern theory and provides some most interesting remarks about these Australians, the latter whom he knew personally, including the comment, "It now seems clear to me, from what I know of these men, that the basic impetus and physical ideas in Green and Ampt (1911) were due to Heber Green; that the large body of meticulous experimentation was due to Gussy Ampt; and that the differential equation involved was set up and solved by my old friend Bumble." The latter is R. J. A. Barnard (1865–1945), a mathematical colleague of Green and Ampt at Melbourne University, where Philip knew Ampt and Barnard when he was a student at this university in the late 1930s. Green and Ampt acknowledge Barnard's help and the help of T. R. Lyle, Professor of Natural Philosophy at Melbourne Unviersity, in their paper.

Treatment of water flow for the first two decades of the 20th century, apart from the work of Buckingham and Green and Ampt, largely was subjective, with water flow and retention data being presented in empirical form in tables and graphs. Considerable data were collected on moisture distribution in soil profiles and published in state experiment station bulletins as well as in Federal Government publications. Typical of many of the bulletins of the state experiment stations is *"Irrigation Experiments in 1901 (on the College Farm)"* (1902) by John A. Widstoe (1872–1952) and W. W. McLaughlin (1876–1966). This 129-page bulletin is a comprehensive report of geological, soil, and irrigation history of this farm and covers all aspects of water flow and retention in the soil and includes crop composition and yield per unit of water. They assert in this bulletin that "the rate of loss of water from soil varies directly with the initial percent moisture in the soil."

Widstoe was a prominent educator in the intermountain west, having been a director of the Utah Agricultural Experiment Station and later president of the University of Utah. His baccalaureate degree in 1894 was from Harvard and his M.A. and Ph.D. degrees were obtained at the University of Goettingen in Germany. In his later years he was a leader in the Mormon Church and one of its Twelve Apostles. His wisdom in the application of science to practical problems undoubtedly was a strong factor in the early development of agriculture and industry in the intermountain west. His contribution to soil science included two texts, *Dryland Agriculture* in 1911 and *Irrigation Agriculture* in 1914, along with numerous technical papers, some of which were in the field of soil physics (W. Gardner and Widstoe, 1921). The author, as a young man, was acquainted with Dr. Widstoe and was greatly impressed by his extensive knowledge expressed in lectures and in numerous books over a wide range of subjects. His image as a very proper Norwegian with dark, greying hair and a pointed goatee is a lasting one.

Results of an exhaustive survey of profile water content were published in Nebraska in 1913 by W. W. Burr (1880–1963), demonstrating the ineffectiveness of dust mulches and the utility of summer fallowing in Nebraska, work continued at Nebraska by Frederick J. Alway (1874–1959), G. R. McDole, J. C. Russell (1889–1976), V. L. Clark and Burr himself (Russell and Burr, 1925). The work of Alway and his colleagues was widely cited in the 1920s and papers by Alway and Clark (1916) and by Alway and McDole (1917) expressed several facts not commonly known until considerably later, when the importance of the water content-dependent unsaturated hydraulic conductivity was fully understood. These observations indicate that the finer the texture, the slower the initial rise of water in a soil column, but the greater the final distance of rise; the coarser the soil, the longer it takes to reach final equilibration; and the higher initial moisture content, the more rapid is downward flow. Considerable use was made of the Briggs and McLane (1907) moisture equivalent and numerous studies were made. One of the many was that of J. C. Russell and W. W. Burr (1925).

Soil physics in America came to vigorous life rather abruptly in the early years of this century, as reported graphically in the author's historical paper (W. H. Gardner, 1977) referred to earlier. The growth corresponds roughly with the formation of the American Society of Agronomy in 1907 (Lyon, 1933) and the first publication of its *Proceedings* in 1909 (later *The Journal of the American Society of Agronomy* (1913–1948), and *Agronomy Journal* in 1949. The very first issue in 1909 contained a paper by Arthur Gillett McCall (1874–1954) entitled "Instruction in Soil Physics" in which it was indicated that "In 1899 a separate laboratory course in elementary soil physics was established at the Ohio State University." The outline headings for the course were not appreciably different from what might be found today. About one third of the soils papers and 13% of the total papers in the *Proceedings* of the ASA, covering the years from the society's inception in 1907 to 1916, when the first issue of *Soil Science* was published at Rutgers University, dealt with soil physics topics. Four of 32 papers in Volume 1, covering the first half year of *Soil Science*, involved soil physics and its first issue was dedicated to E. W. Hilgard. As indicated earlier, the journal *Hilgardia* also was named in his honor in 1925. The first volume of this journal contained a paper on soil temperature as influenced by a paper mulch (Charles F. Shaw, 1926) and numerous papers of importance in soil physics have appeared over the years since. In 1939 W. P. Kelley published a short biography of Charles F. Shaw.

Wollny's journal, *Forschungen auf dem Gebeite der Agrikultur-Physik*, published annually from 1878 to 1898, assured Wollny the reputation as Germany's foremost soil physicist. The British *Journal of Agricultural Science*, published in Cambridge beginning in about 1865, and the *Discussions of the Faraday Society*, founded in 1903, were major sources of soils information in England, as were papers in the American Bulletin Series, begun in 1895 by the USDA, Division of Agricultural Soils (becoming the Bureau of Soils), and later the *Journal of Agricultural Research*, published by the USDA from 1914 through 1949. The *Canadian Journal of Soil Science*, begun in 1920; the *Journal of Soil Science*, begun at Rothamsted in 1949 as the journal of the newly formed (1947) British Society of Soil Science (*Imperial Bureau of Soil Science, Technical Communications*, published out of the Rothamsted Experiment Station from the 1920s through most of the 1940s preceded this journal); the *Netherlands Journal of Agricultural Science*, begun in 1953; the Japanese

Soil Science and Plant Nutrition, begun in 1955 as *Soil Science and Plant Food; Soviet Soil Science (Pochvovedenic)* , begun in 1899 with an English translation in 1958; the *Australian Journal of Soil Research*, begun in 1963; and *Geoderma*, an international journal begun in 1967 and published in The Netherlands with an international editorial board are the major journals cited currently in soil physics research papers. However, there are numerous other journals, such as the *Transactions of the American Geophysical Union*, begun in 1920 and expanded into several other journals, including *Water Resources Research* in 1965. The most widely used abstracting journal in the English language, *Soils and Fertilizers*, Harpenden, England, was begun in 1938 as an offspring of an earlier series, *Publications Relating to Soils and Fertilizers*. Important papers have appeared in other scientific journals from time to time, such as *Plant Physiology*, in which the first volume in 1926 contained three papers on soil physics, and the *Journal of the American Society of Agricultural Engineers* first published in 1920, or in its *Transactions*, begun in 1958. The *Physical Review* has contained several early papers of importance to soil physics and *Science* and *Nature* in England and *Science* in the United States frequently carry articles involving soil physics.

Soil physics in the United States and in other countries received a boost with the formation of the International Society of Soil Science in Rome in 1924 (Lipman, 1928) with its first congress, held at Washington, D.C. in 1927, being opened by United States President Calvin Coolidge. A Hungarian, Alexius A. J. De'sigmond, discusses the formation of the ISSS in a 1935 paper. Of some significance to soil physics history is the fact that Soil Mechanics and Physics was the first of six commissions formed with Dr. Novák of Czechoslovakia as president (Keen, 1928a). Twenty-two papers were presented at the first meeting, with approximately one third of these by American soil physicists, including two, H. E. Middleton (1894–) and L. B. Olmstead (1884–1952), who would later become S-1 Division chairmen in the SSSA. Eleven of the 22 papers presented covered mechanical analyses, seven water, two draught measurements on plows, and one each organic matter and temperature. Middleton (1920) earlier had published a paper in which he had related the moisture equivalent to the mechanical analysis of a soil.

The first American organization of soil physicists was created, with the inauguration of the Soil Science Society of America in 1936, as Division S-1, Soil Physics. The parent societies of the SSSA were the Soils Section of the ASA and the American Soil Survey Associaton. The divisions were patterned after those of the International Society of Soil Science. Leonard D. Baver (1901–1980) was the chairman of the subsection on soil physics under the Soils Section of the ASA at the time of organization and may be considered to be the first chairman of Division S-1. Over the years numerous soil physicists have held the office of president of the SSSA and the ASA. J. F. Lutz (1907–), the 1940 chairman of Division S-1, in

1977 prepared an excellent "History of the Soil Science Society of America" (Lutz, 1977). It was in this era that the Soil Conservation Service was created in the United States to deal formally with preservation of soils, which were being lost to wind and water erosion. Hugh Hammond Bennett (1881–1960), whose work and writings (1939; Brink, 1947), so clearly demonstrated the need for conservation, was a major force in this development.

George J. Bouyoucos (1890–1981; Figure 11) of the Michigan Agricultural College, began contributing to soil physics in his early 20's. Bouyoucos, born at Likohia, Greece, at age 11 left his father's sheep herds and traveled by himself to Chicago. After attending a YMCA school, where he learned English, he entered the University of Illinois at age 15, graduated 3 years later, and entered Cornell University, where he completed his Ph.D. in 1911 at the age of 21 and joined the staff at Michigan. Bouyoucos' early work involved temperature and included some work in the effect of temperature on movement of water vapor and capillary moisture. His best known contributions to soil physics are his widely used gypsum moisture blocks, (Bouyoucos, 1947) which bear his name and the hydrometer method for particle size analyses (Bouyoucos, 1927a,b). His work continued well up until his death in 1981. The form of water in the soil still was a matter for much study in the early part of the 20th century. Bouyoucos (1921) separated water into three categories: gravitational water, which moves readily under the influence of gravity; free water, which freezes at −1.5°C and is available to plants; and unfree water, which is held in capillaries and is only slightly available to plants. The unfree water category included also "combined" or hygroscopic water, which is not available to plants. In 1917, he had published evidence, obtained using dilatometry, which led him to believe that the hygroscopic film might be in a soild phase and, possibly, in chemical

Figure 11. Bouyoucos, George J., (1890–1981). Courtesy of American Society of Agronomy.

combination with soil minerals. In any event the hygroscopic water was generally regarded as that part of the water that did not participate in capillary or gravity flow. Its determination involved measuring the water content of soil in equilibrium with the atmosphere or with atmospheres of varying degrees of moistness extending up to saturation.

Following up on earlier work of Mitscherlich and others, in a Bureau of Soils bulletin in 1908 H. E. Patten (1873-?) and F. E. Gallagher (1884-1950) reported that the amount of water vapor absorbed by a soil from a saturated atmosphere decreased as the temperature increased. Patten (1908) also dealt with heat transfer in soil. Several years later, in 1921, and in 1924, Moyer Delwin Thomas (1894-1975) associated the capillary potential with vapor pressure and in 1926 Leon B. Linford (1904-1957) in a masters thesis under Willard Gardner (1883-1964), showed experimentally and theoretically, using potential theory, that the amount of water absorbed in a soil matrix depended on the vapor pressure and that an equilibrium value other than saturation could not exist in a saturated atmosphere, provided that radiant energy was not permitted to raise the temperature of the soil above that of the surrounding atmosphere. Another Gardner student, Scott Ewing, in 1922 measured the rate of flow of saturated water vapor through quartz flour. In 1912 Canadian C. J. Lynde described soil as acting like a semipermeable membrane and, with F. W. Bates, Lynde (1912) reported "Further studies in the osmosis of soils." In 1913 Lynde and H. A. Dupre reported "On a new method of measuring the capillary lift of soils." Capillary rise of water was still an important topic two decades later when H. A. Wadsworth described the nature of capillary rise (1931); W. O. Smith, writing in *Physics*, dealt with the "minimum capillary rise in an ideal soil" (1933); and 3 years later in *Soil Science*, the "sorption in an ideal soil" (1936). Ten years earlier H. A. Fisher (1926) had described "the capillary forces in an ideal soil." In 1939 Smith also dealt with thermal conductives of moist soils.

VIII. The Beginning of the Modern Era

The distinction between gravity and capillary water was not clear in the early decades of the 20th century. Briggs' (1897) and Bouyoucos' (1921) classifications were followed by somewhat comparable classifications by Aleksander Feodorovich Lebedeff (1882-1936) in 1927 and F. Zunker in 1930. Lebedeff separated water into four forms: water vapor; hygroscopic water, which is held on particle surfaces by forces of adhesion; film water, held in pores by virtue of molecular forces of cohesion; and gravitational water. Zunker had seven classes of water :"osmotic water," associated with organic materials; hygroscopic water, held on particle surfaces by adhesion; capillary water, held by capillary forces in pores connected with free water; "held" water, retained by surface tension forces in films,

Figure 12. Gardner, Willard, (1883–1964). Author's father, family photograph.

pore angles, and capillaries not connected with ground water; gravitational water that moves in capillary pores by gravitational forces and in pores larger than capillary size; ground water that exists in "tension-free" pores spaces; and water vapor. However, King in 1901 had shown that a soil column, allowed to drain in a gravity field, was still undergoing some readjustment after a period of several years. This, of course, made any classification, as a static property, questionable. The true dynamic nature of the water retention problem was not clearly recognized until much later.

A. N. Kostiakov was an important early contributor to water-flow theory. One of his many papers dealt with the dynamics of water percolation, indicating the necessity for taking a dynamic point of view (1932). Mortimer R. Lewis (1886–?) was an early contributor to irrigation practice, one of his papers dealing with infiltration rate (1937). Infiltration also was the special interest of Robert I. Horton, who published numerous papers on the subject, including one discussing an approach to a physical interpretation of infiltration capacity (1940).

Although the Briggs and McLane moisture equivalent (1907) was widely used as a measure of the capacity of a soil to retain water that could be used by plants, uneasiness over such a concept developed. Willard Gardner (1919, 1920) (Figure 12) indicated that if the centrifugal force used by Briggs and McLane to bring soil to the "moisture equivalent" was proportional to Buckingham's "capillary potential," then the capillary potential likely was a hyperbolic function of the moisture content, at least over the most practical part of the range. He showed hypothetical curves for the relationship between capillary potential and moisture content for different values of particle radius which would be an index of capillary pore size. Later he and several co-workers (W. Gardner *et al.*, 1922) presented such an equation for one of their experiment station soils, $(\psi + 145)(-0.173) = -42.6$. Vertical lines of constant potential were drawn on the hypothetical moisture content-capillary potential curves to represent various soil-moisture "constants," such as "hygroscopic moisture," "moisture equivalent," and the "wilting coefficient" of Briggs and co-workers. The arbitrary nature of such constants was thus demonstrated; but such constants did serve many practical purposes and have been widely used, persisting even today.

The concept of *field capacity*, which has evolved from a background of "soil moisture constants," was rejected by Gardner, Richards, and others and it continues to be ambiguous. Its dependence on wetting history and unsaturated conductivity, which may differ appreciably at various depths in the soil profile, appears not always to be recognized [E. A. Coleman (1910–?), 1944; L. A. Richards, 1950].

Expansion of Buckingham's ideas on moisture flow was stimulated with the development of the tensiometer by Willard Gardner and his associates in 1922. Although Buckingham had applied the term "capillary potential" to the energy state of water, his only way of measuring it involved equilibration with gravity in a tall column. Gardner interposed a porous cup with pores fine enough to remain water filled over a considerable range between the soil and a water manometer. At equilibrium the negative pressure indicated by the manometer was taken to be the tension in the soil water, or an index of the capillary potential. Experimental development of this idea in Gardner's laboratory, as attributed to Gardner and described by O.W. Israelsen (1887–1968) in 1927 and later expanded by Lorenzo A. Richards (1904–) and Willard Gardner (1936), constituted the well-known tensiometer, which now is the basic instrument for measurement of matric potential in the wet range above −1 bar. L. A. Richards' extension of the idea to the pressure membrane and porous plate apparatus for making measurements beyond the 1-bar range (1941) constitutes a major contribution to modern soil water research.

Gardner was born in southern Utah. Educated initially for a career in commerce, he started out as a bank clerk and stenographer. However, his interest in mathematics and physics led him to leave banking to pursue a B.S. degree in physics while earning his way as a stenographer-clerk in the Utah Agricultural Experiment Station. After completing his B.S. degree in 1912 he went to the University of California as a teaching assistant and later instructor in physics, earning the Ph.D. in 1916 with a thesis on the photoelectric effect. He was chairman of Division S-1 of the SSSA in 1941 (S. A. Taylor, 1965). It is of interest to note that, as has been true of numerous early as well as some more modern scientists who have come to be known as soil physicists, Gardner's educational background is not soils, soil water having become an acquired interest. At the University of California he was friendly with O. W. Israelsen, who was studying agricultural engineering on leave from Utah Agricultural

College. This friendship undoubtedly had some influence on his interest in soils and his later appointment to the faculty of Utah Agricultural College as physicist in the Experiment Station, where he had earlier served as a clerk. In the intervening period he served as principal of an academy and as an instructor in physics at Brigham Young College.

Gardner was particularly noted in his early professional years for use of mathematics not well understood by many of his contemporaries, particularly by agricultural engineers. Some of them chided him for not writing his papers more simply so that they would be better understood. Ren (Lorenzo) Richards has told a story on tape (copies retained by the author and by Wilford R. Gardner) of a visit to Gardner's laboratory by Sir Bernard Keen in about 1925. Richards, a graduate student at the time, was in an adjacent room partitioned off only with glass so that the conversation was easily witnessed. Sir Bernard took Gardner to task because his papers were so "hard to read." "Dr. Gardner was a man of some piety and had a quiet manner of speaking, but he had a prominent lower jaw. I could feel the air crackle in the next room as he paused for reply, which was, 'Sir Bernard, God made soil physics hard, not Willard Gardner.' "

The era of the 1920s produced a number of Untied States soil physicists who have played important roles in the ongoing of the science and of professional soils organizations. Among these are Niels E. Edlefsen, Richard Bradfield, Leonard D. Baver, Geoffrey B. Bodman, Frank J. Veihmeyer, and Lorenzo A. Richards. Some of these poeple, along with Gardner and a Brigham Young University chemist, Thomas L. Martin (1885–1958) (father of soil scientists William P. and James P.), are educational progenitors of numerous present-day American soil physicists.

Niels E. Edlefsen (1893–1974) was educated in physics at Utah State University. After receipt of the bachelor's degree in 1916 he taught briefly at Utah State over two periods of time and at the University of California, ultimately receiving his Ph.D. from there in 1930. He had a brief career on the California Experiment Station staff until 1941, when he became engaged in war-related research at the Massachusetts Institute of Technology. From there he went on to technical and management positions in the government and industry. Edlefsen was interested in vapor pressure in soil and in the broad field of thermodynamics of soil moisture. He collaborated with Alfred B. C. Anderson (1906–) in writing a comprehensive monograph "Thermodynamics of Soil Moisture" (Edlefsen and Anderson, 1943), published as an issue of *Hilgardia*, containing 175 references, and which was for a great many years the standard reference in this area of soils. Anderson received his Ph.D. in physics in 1934 from the California Institute of Technology and worked for a short period as a soil physicist with the U.S. Department of Agriculture, after which he went into wartime research and industrial work. In addition to

Figure 13. Bradfield, Richard, (1896–1981). Courtesy of American Society of Agronomy.

collaborating with Edlefsen on soil moisture thermodynamics, he also worked on a method for using electrical capacitance to measure soil moisture (Anderson, 1943).

Richard Bradfield (1896–1981; Figure 13) was born and educated in Ohio with a B.A. degree from Otterbein College in 1917 and his Ph.D. from Ohio State University in 1922. Both Bradfield and Baver had early associations with G. W. Conrey (1887–1948) at Ohio State University (Mason, 1948). Bradfield was on the faculty at the University of Missouri from 1922 to 1930, Ohio State University from 1930 to 1937, and Cornell University from 1937 to 1962, where he was department head until 1955. He became Emeritus Professor in 1962 and then undertook several assignments in the far east, culminating as Director of the International Rice Institute under the Rockefeller Foundation, where he served until the mid 1970s (Cline, 1973). Bradfield's contributions were more in soil chemistry than in physics. However, his extensive work with colloidal clay relates strongly to and has considerable importance to soil structure as studied by the soil physicist (Bradfield, 1925; Bradfield and V. C. Jamison, 1939). He was the first president of the SSSA in 1937, ASA president in 1942, and ISSS president from 1956 to 1960.

Leonard D. Baver (Figure 14) was born in Ohio in 1901 and received his B.S. and M.S. degrees at Ohio State University in 1923 and 1926, and his Ph.D. from the University of Missouri in 1929. Following graduation he started his soil science career at Alabama, serving there for 2 years before returning to the University of Missouri in 1931. In 1935–1936 he served as Senior Soil Conservationist with the Soil Conservation Service and in 1937 joined the faculty at Ohio State University for a 3-year period before becoming Agronomy chairman at North Carolina State College in 1940, Director of the Agricultural Experiment Station in 1941, and Dean and Director in 1942. In 1948 he became Director of the Hawaiian Sugar

Figure 14. Baver, Leonard D. (1901–1980). Courtesy of American Society of Agronomy.

Planters Association. He returned to Ohio State again in 1965, from which he retired as Emeritus Professor in 1972, with a 1-year interim assignment as chief of Party, USAID, in India. Baver always was much interested in the professional development of soil science and, in addition to being the first chairman of the soil physics division in the SSSA, during World War II he served a 2-year term as SSSA president. He was a capable administrator and prior to his moving to Hawaii he was considered for, and possibly invited to become, president of Ohio State University. In a reception for him on his arrival in Hawaii he was asked if he might return to Ohio State as president, to which he responded "I should spend my life worrying about the success of the football team and the morals of the coeds?"

Geoffrey B. Bodman (1894) was born in England but came to Canada at an early age. He obtained his B.S.A. degree from the University of Saskatchewan College of Agriculture. After instructing in science at the School of Agriculture in Alberta he entered the graduate school at Minnesota, where he earned the M.S. and Ph.D. degrees in 1924 and 1927. He joined the faculty at the University of California in 1927, retiring to Emeritus status in 1962, having srved as department chairman for 7 years, as a soil scientist with the USGS and Corps of Engineers on war duty from 1943 to 1945, and as a visiting professor in Taiwan and later in Egypt. Bodman was president of the SSSA in 1958. His interests ranged widely across the field of soil physics, particularly in soil moisture and its energy state (Bodman and Edlefsen, 1934; Bodman and Coleman, 1943).

Frank J. Veihmeyer (1886–1977) in the late 1920s, with colleagues, made extensive studies of soil water conditions and plant growth (Conrad and Veihmeyer, (1929) that later evolved into a definition of available water, which was the difference between field capacity (not rigorously defined) and wilting point (Veihmeyer and Hendrickson, 1931, 1933). Their extensive studies on soft tree fruits and other crops led them to conclude "that plants can use water with equal facility throughout the range of moisture contents from the field capacity nearly to the permanent wilting percentage." Their observations and their concept of available water became widely accepted and, although not supported by modern water-plant growth evidence, did serve to make irrigation a quantitative science. With modifications relative to what part of the available water should be used prior to irrigation for optimum crop growth, it is widely used today. Some years later the matter was put into much better perspective by L. A. Richards and Wadleigh (1952). Veihmeyer also collaborated with N. E. Edlefsen on at least one occasion (Veihmeyer 1943). He was born in Washington, D.C. and received his B.S. and C.E. degrees at George Washington University in 1912 and 1913. He obtained his Ph.D. from Johns Hopkins University in 1927 and began his professional career with the USDA in 1913 after having worked as an assistant in the USDA while pursuing his B.S. and M.S. degrees. He joined the Division of Irrigation of the College of Agriculture at the University of California in 1918, from which he retired to Emeritus status in 1955, having served as division chairman for 18 years having fulfilled numerous international assignments. He continued professional work as an Emeritus Professor for many years.

Ross Edgar Moore (1898–1958) a short-time contemporary of Bodman and Veihmeyer, at the University of California, contributed appreciably to the understanding of water flow and retention in agricultural soils as illustrated in a noted *Hilgardia* publication on water conduction from shallow water tables (Moore, 1939). He also reported work on the relationship of soil temperature to water retention and infiltration (Moore, 1941). In 1939 Frank L. Duley (1888–1978) in Missouri reported on the influence of surface factors on water intake.

Early work on water flow also came out of Austria, where Karl Terzaghi (1883–1963) wrote (1923) a diffusion equation to describe horizontal, one-dimensional, saturated flow in swelling soils, which Philip connects with Buckingham 45 years later (Philip, 1974). Terzaghi later joined the faculty of the Massachusetts Institute of Technology in Boston, where he was well known for his work in soil mechanics (Terzaghi and Peck, 1948). Another contributing scientist from outside of the immediate field of soils is Irving Langmuir (1881–1957), whose work on adsorption of gases and general physical chemistry supplied needed information and experimental techniques for study of colloidal soil (Langmuir, 1918). Arthur W. Adamson's 1960 book *Physical Chemistry of Surfaces*, particularly Chapter I, "Capillarity," also has been useful.

L. A. Richards (Figure 15) entered Utah Agricultural College a few years after Gardner had joined the faculty, completing his B.S. degree in 1926 and an M.S. in 1927. From there he went to Cornell University,

Figure 15. Richards, Lorenzo A., (1904–).

where he received his Ph.D. in 1931 and served as an instructor in physics until 1935. After brief employment at the Battelle Memorial Institute, he joined the faculty of the Iowa State University, where he served until 1939 when he became soil physicist at the newly founded U.S. Department of Agriculture Salinity Laboratory at Riverside, California. Except for two years at California Institute of Technology working on rocket ordinance during World War II, he remained at the Salinity Laboratory until formal retirement in 1966. He has continued his work in soil-physics-related research for many years since his retirement. His career and contributions to soil physics are well documented in a 1972 issue of *Soil Science* dedicated to him (Bower, 1972).

Richards was chairman of Division S-1, SSSA in 1938, SSSA president in 1952, and ASA president in 1965. In addition was his extensive work on instrumentation, which included tensiometers, pressure membrane, and porous plate apparatus for production of desorption curves, modulus of rupture apparatus, and a psychrometric method for measuring vapor pressure in soils (L. A. Richards, 1941, 1942, 1948, 1949, 1953; L. A. Richards and Fireman, 1943; L. A. Richards and Ogata, 1958, 1961).

In two papers, Richards (1928, 1931) laid the foundation for much of the work on unsaturated flow that has followed. He was the first to apply continuity to the Darcy equation to produce the important nonlinear differential equation for unsaturated flow. The 1928 paper, covering work at the Utah station and written as he began doctoral studies at Cornell University, among numerous other important ideas, contained a succinct statement about availability of water to plants: "The term 'availability' involves two notions, namely, (a) the ability of the plant root to absorb and use the water with which it is in contact, and (b) the readiness or velocity with which the soil water moves in to replace that which has been used by the plant." This idea, forgotten or overlooked and then rediscovered more than once, cannot be improved upon today. As

Wilford R. Gardner (1925–) says in a 1972 paper in the *Soil Science* issue honoring L. A. Richards, "It is doubtful whether any other soil physicist, before or since, so nearly succeeded in writing a charter for his entire research career at the very outset as did L. A. Richards in his first publication."

Many of Richards' research contributions were achieved while he was on the staff of the United States Salinity Laboratory in Riverside, California. Here he collaborated with numerous soil physicists and, particularly, with other soil and plant scientists with related interests involving their own special fields. In 1943 Richards and L. R. Weaver reported work showing that the water content of soil at 15 bars tension corresponded closely to that in soil when plants wilted permanently in accordance with the definition given by Briggs and Shantz in 1912. A particularly important contribution was with Cecil Wadleigh (1907–), a plant physiologist colleague at the Salinity Laboratory, with whom he collaborated on a chapter, "Soil Water and Plant Growth," in 1952 monograph, *Soil Physical Conditions and Plant Growth*, edited by Byron T. Shaw (1907–), another soil physicist of note in this era. Shaw is particularly noted for his use of the electron microscope to reveal the shape of clay particles (B. T. Shaw, 1942). Wadleigh's presence on the Salinity Laboratory staff signaled an important advance in this laboratory's soil water–plant research program by bringing together soil and plant physics. Typical of his contribution was a study of leaf elongation, which showed the leaf's strong dependence on soil moisture stress (Wadleigh and Gaugh, 1948). Another plant physiologist, Paul J. Kramer (1904–) at Duke University, contributed similarly with a paper on water absorption by roots (P. J. Kramer, 1932) and with a book, (P. J. Kramer 1949) *Plant and Soil Water Relationships*. A major achievement of the Salinity Laboratory was Handbook 60 (U.S. Salinity Laboratory, 1954), initiated and edited by Richards and containing, among other things, state-of-the-art information on soil physics and measurements of physical properties and processs. This handbook became an essential tool of soils laboratories throughout the world.

In 1960 Richards gave a general lecture on "Advances in Soil Physics," representing Commission I at the 7th International Congress of Soil Science, Madison, Wisconsin (L. A. Richards, 1960). In this paper he reviewed much of the work on water flow and retention since Buckingham's 1907 paper. He indicated of Buckingham: "His ideas were so clearly expressed, his terminology so apt, and his analyses so sound, I cannot avoid the feeling that during the last 50 years of work on the physics of soil water, we have mainly been filling in between his lines." He refers to Buckingham's capillary conductivity function and his own 1931 experimental measurements of this in the tensiometer range, and to the extension of such measurements to the entire plant growth range using a membrane outflow measurement by W. R. Gardner in 1956 and "verified and refined" in 1958 by E. E. Miller (1915–) and D. E. Elrick (1931–) and by P. E. Rijtema in 1959. He refers also to the field capacity concept, indicating that "It is the author's prejudice that the concept of field capacity has done more harm than good" (L. A. Richards, 1955). Richards collaborated with D. R. Neal and M. B. Russell on a paper "Observations on Moisture Conditions in lysimeters" (Neal, *et al.*, 1937)

Richards engaged formally in the education of soil physicists only at Iowa State where Morell Belote Russell (1914–) received his Ph.D. in 1939 with him. However, the U.S. Salinity Laboratory cooperated in graduate programs with the University of California at Riverside so that his influence on education of students in soil physics continued. Russell's Ph.D. thesis was on heat of wetting in soils (M. B. Russell and Richards, 1938). He remained in teaching and research at Iowa State until 1945, when he joined the faculty at Cornell University. There he was associated with a number of Ph.D. students in soil physics, incidentally linking Richards educationally to a very long line of soil physicists who were at Cornell with Russell. These included Arnold Klute (1921–), Sterling A. Taylor (1918–1967), and Robert D. Miller (1919–), all of whom have had outstanding Ph.D. students working with them. R. D. Miller is the son of Merritt F. Miller (1875–1965) who, with Frank L. Duley, established the Miller-Duley Erosion Plots at the University of Missouri-Columbia Campus, and brother of E. E. Miller physicist–soil physicist at the University of Wisconsin. In 1959 M. B. Russell went to the University of Illinois as department chairman and in 1962 he became the Director of the Illinois Agricultural Experiment Station. Since 1973 he has been involved in international agricultural programs. Russell was the President of Commission I, Soil Physics, of the International Society of Soil Science from 1956 to 1960 and became Chairman of SSSA Division S-1 in 1948, President in 1955, and President of the American Society of Agronomy in 1963. His research contributions have included work in soil water–plant relations (1960), soil structure, and aeration (M. B. Russell, 1949, 1959) and he wrote the chapter on aeration (1952) in the monograph *Soil Physical Conditions and Plant Growth* (B. T. Shaw, ed). Russell also gave strong support to use of an energy concept of soil moisture (1943)

The concept of water potential and its various components, under various names, such as tension, capillary tension, suction, and so forth, became of increasing importance as its measurement improved. In 1937, Robert Gardner (1888–1977), brother of Willard Gardner and father of two later soil physicists Wilford R. (1925–) and Herbert R. Gardner (1928–), described a method for measuring the capillary tension of soil moisture over a wide moisture range. Sterling J. Richards (1909–1979), in 1938 described the calculation of soil moisture from capillary tension records, and Paul R. Day (1912–) in 1942 discussed the moisture potential of soils. In 1943, Hans F. Winterkorn (1905–) published a paper on the condition of water in porous systems. Helmut Kohnke (1901–) dealt with

Figure 16. Kirkham, Don, (1908–). Photograph courtesy of American Society of Agronomy.

the practical use of the energy concept in 1946 and later, in 1968, he wrote a text on soil physics.

Sterling J. Richards, half-brother and contemporary of L. A. Richards and also from the same Utah school, was involved in soil physics research at Rutgers University and later at the University of California at Riverside. He worked extensively on soil water, and at Riverside, particularly, on application of soil water physics to practical problems of plant growth. Some of this work was with A. W. March (1913–) who, as an expert on soil water physics, had considerable influence on irrigation practice in Oregon and California.

L. A. Richard's successor at Iowa State at the cessation of World War II was Don Kirkham (Figure 16). Kirkham (1908–) is another soil physicist who came under the influence of Willard Gardner at Utah. In 1938 he came to the Utah state Agricultural College as an instructor in physics from Columbia University, where he had obtained his Ph.D. degree with a thesis on paramagnetism. He became interested in drainage research there and found his physical-mathematical background ideal for dealing with underground water flow. His first work was to make a dye-tracer model showing the streamlines as water moved toward drains, confirming the curves produced theoretically some years earlier by Willard Gardner. One of this models showed two drains, one in a clay soil and an adjacent one in a coarser soil. As predicted by theory—but not believed by some drainage engineers, one of whom was the Dean of Engineering—the position of the streamlines was unaffected by the permeability of the soil materials (Kirkham, 1940). The author was present in the laboratory when the Dean came to see the flow model and recalls him scratching his head and remarking that he would need to go back to his office to think this over. Kirkham's work on saturated flow and drainage systems. together with excursions above the water table and into other soil physics problems such as soil aeration (Evans and Kirkham, 1949), with numerous collaborating students, has made him probably the outstanding soil physicist working on saturated water flow in his era. His book with W. C. Powers, *Advanced Soil Physics* (1972), and numerous papers with his many students, including such names as J. N. Luthin (1915–1981), W. R. Gardner (1925–), C. H. M. Van Bavel (1921–), D. R. Nielsen (1931), and Dale Swartzendruber (1925–) have greatly enhanced the science.

Kirkham taught physics at the Utah school until he was called into World War II naval research at the Massachusetts Institute of Technology, where he worked on sonar underwater detection systems. Following the war years he went to Iowa State Unviersity in the position in physics and agronomy occupied earlier by L. A. Richards. Kirkham's review papers on soil physics (1961) physics (1961) and on drainage reserach (1972) provide useful insights into the growth of soil physics. A 1978 review of Kirkham's work by one of his students, Dale Swartzendruber (1925–), shows Kirkham's numerous contributions. He is known by his many students for his practice of using advanced students in guiding newer students in their studies and research.

Several English soil physicists, including Sir Bernard A. Keen (1890–1981; Figure 17), E. C. Childs, R. K. Schofield, William B. Haines, Gilbert Wooding Robinson, H. L. Penman, and the Indian, A. N Puri (1898–1971), were contemporaries of Richards and contributed materially to the growth of soil physics.

Bernard Keen's major contribution appears to have been his book, *The Physical Properties of the Soils* (1931), in which he not only discusses his own work but reviews the history of soil physics, providing a highly comprehensive bibliography not available before. He worked at many

Figure 17. Keen, Sir Bernard, (1890–1981). From photo on page 21, Vol. 25, *Soil Science*, 1928.

things, including an early attempt to describe the energy status of the soil solution using freezing-point depression (Keen, 1919). He considered evaporation from soil (1914), dealt with factors determining soil temperature with (E. J. Russell, 1921), and discussed moisture relationships in an ideal soil (1924) and the significance of "single value soil properties" (with J. R. Coutts, 1928). His interests extended well beyond soil moisture; he reported on the relation of clay content and certain physical properties of a soil (with H. Raczkowski, 1921), and discussed national and international use of soil mechanical analysis (1928b).

In 1925 Keen gave a lecture on the place of physics in the theory of agriculture ("The physicist in agriculture," Keen, 1926) in the rooms of the Chemical Society, London. His discourse was introduced by Sir Daniel Hall, Chief Scientific Adviser to the British Board of Agriculture and Fisheries, who talked about "a new field of investigation that has been disclosed of late years." From the context of the early part of his talk it appears evident that he was talking to his "fellow physicists."

The lecture was long and comprehensive—32 pages, some 15,000 words—covering plant-environment relationships, water-flow theory, colloidal properties of soil, and extending through cultivation and the implements used. He discusses "Poiseuille's capillary tube law" and a study of pore space in an elaboate system of spheres by Slichter in 1890, "who arrived at formulae for the flow of water through it." He mentions a parallel experimental investigation by King and the work of Green and Ampt who, "some ten years ago," reinvestigated the problem in Australia. It is of interest that Keen nowhere mentions the work of Buckingham in this lecture but does include Buckingham in his 1931 book. Also, although he mentions the work of Slichter here he makes no reference to him in his book. Toward the end of his lecture Keen shows a map of a portion of a Rothamsted experimental field, which on the surface appears to be perfectly uniform, and upon which lines of equal drawbar pull, "isodynes," are drawn. His remarks are directed toward problems associated with use of such heterogenous fields for research involving mechanical properties of soil. Field heterogeneity has only in the last decade received careful study by soil physicists.

In the opening sentence of the lecture Keen says, "Agriculture may be defined as a process of interference with Nature." He then goes on to say that "If this interference is to be systematic—and the increasing food requirements of the world demand that it shall be systematic—it follows automatically that science can provide useful and probably indispensable aid." He then indicates that he is limiting his discussion to the application of physics to soil problems, his reason of greatest influence being his "desire to show my fellow-physicists that the soil itself is a most attractive study, presenting problems whose variety and complexity will satisfy—and at times alarm—the most enthusiastic research physicist."

In England G. W. Robinson (Biography in Muir, 1950) was interested in the mineral composition of soil and developed a method for mechanical analysis (1922) and for the representation of data in mechanical composition curves (1924). His 1932 book, *Soils: Their Origin, Constitution and Classification*, deals extensively with such physical properties of soils. E. W. Russell, son of Sir E. John Russell, worked extensively on soil structure and in 1934 reported work on the interaction of clay and organic liquids with soil crumb formation. He later reported on the physical basis for soil structure (1938a) and in the same year he wrote a book on soil structure (1938b).

In the United States Robert E. Yoder reported a direct method for soil aggregate analysis involving wet sieving and described the physical nature of erosional losses (1936). George M. Browning (1937) reported on changes in errodibility brought about by application of organic matter. J. R. McHenry and M. B. Russell dealt with the mechanics of aggregation of puddled materials (1943). In 1944 Thomas M. McCalla (1909–) reported on a water drop method for determining the stability of soil structure. W. S. Chepil in 1943 related wind erosion to water-stable dry clod structure and did considerable work following World War II on factors influencing clod structure and wind erosion (1951). Also, in the postwar years A. W. Zingg carried out field studies involving use of a portable wind tunnel (1953).

Rothamsted Laboratory soil physicists, R. K. Schofield (1901–1960) and H. L. Penman (1909–1984), contributed appreciably to soil physics in the decades of the 1930s and 1940s. Schofield is noted particularly for his introduction of the pF of soil moisture, of which he says in his 1935 paper, "The pF is the logarithm of Buckingham's potential. By analogy with Sorensen's acidity scale the symbol *p* indicates its logarithmic character, while the symbol *F* is intended to remind us that by defining pF as the logarithm of the height in centimeters of the water column needed to give the suction in question, we are really using the logarithm of a free energy difference measured on a gravity scale." Of his omission of capillary from Buckingham's "capillary potential" he notes that he has "deliberately not used this term because the word capillary brings to so many minds thoughts about surface tension." It is interesting to note that he refers to the 1-atm limit of a Buchner funnel "so that pF 3 marks the limit for such experiments as have already been described. Higher values could presumably be reached by using air pressure in conjunction with a suitable filter, but in the absence of any measurements of this kind they must be investigated by other means." Evidently he was unaware of the pressure-membrane apparatus used by L. A. Richards (1941) earlier. He then remarks that "freezing point and vapour pressure are of use here." In his pF paper Schofield discusses a number of problems associated with water flow and retention that would be identified today as having to do with hysteresis or with the often dominating character of low values of the water-content-dependent unsaturated conductivity. In this connection Schofield makes, to the author at least, one very profound observation, taken slightly out of context, "that the soil moisture conditions which are of most practical interest and importance are not conditions of true equilibrium."

Schofield was interested in a wide range of physical problems (e.g., Schofield, 1935; 1943; 1946; 1950) and he sometimes collaborated with H. L. Penman (Schofield and Penman, 1949). H. L. Penman is most noted for his work on evaporation from soil and plant surfaces and for the equation that carries his name (Penman, 1948a,b). This equation describes transpiration in terms of environmental parameters and diffusion resistances. Penman was the director of the Rothamsted Laboratories for a time and lectured widely on applications of physics in agriculture as is indicated by his many papers, some of which are a given as references here (Penman, 1940; 1948b; 1949a, b). In at least one paper Penman reversed the usual agricultural interest on the influence of environment on transpiration and dealt with the role of vegetation on meteorology (transpiration), soil mechanics, and hydrology (1951). In the United States in about the same period, C. W. Thornthwaite (1899–) and B. Holzman worked on similar problems and developed equations for evapotranspiration. A USDA bulletin describes some of their work (Thornthwaite and Holzman, 1942).

Ernest Carr Childs (1907–1973; Figure 18) born in East London, received his advanced education at Cambridge in physics with Ph.D. degrees from King's College in 1931, with a thesis on the radiofrequency properties of ionized air, and from the University of Cambridge in 1934, with a thesis on the diffraction of slow-speed electrons in certain metal vapors. Childs' entrance into the study of soils came when he accepted the post of physicist in the Unviersity of Cambridge School of Agriculture and began the study of land drainage.

Childs pioneered the use of electric analogs in drainage studies (1936), later extending his interests into the properties of porous materials and flow of fluids through them. His sand-tank model, called "the Tank" by the many students and visiting scientists who used it in the study of flow and water-table behavior in drainage situations, is well known. In the late 1940s he put Darcy's law into the form of a diffusion equation with a moisture-content-dependent diffusivity. Childs was an early contributor to modern unsaturated flow theory (Childs and N. Collis-George , 1950a, b). He was one of the first to calculate hydraulic conductivity of a porous material from the moisture characteristic curve. This work is discussed by W. R. Gardner (1974) in an issue of *Soil Science* dedicated to Childs, with an introduction written by several of his students (Youngs et al. 1974) and with papers by students and colleagues who had worked with him [Collis-George (1925–), 1974; W. R. Gardner, 1974; Marei, 1974; Poulo-

Figure 18. Childs, Ernest C., (1907–1973).
From Vol. 117, Soil Science, 1973.

vassilis and Tzimas, 1974; Smiles, 1974; Youngs, 1974], particularly on visits to his Cambridge laboratories for various periods of time. Philip captures the essence of Childs' character in his paper (Philip, 1974b) where he says: "Ernest Carr Childs was one of the small number of intellectual leaders of soil physics over the last forty years. In his forthright and literate way, he insisted on clear thinking and decent rigor in a field where scientific standards have seemed, all too often, to receive little consideration."

Childs was particularly interested in the phenomenon of pore-water hysteresis and worked the last few months of his life on a "unified domain theory," unfinished at the time of his death. Numerous papers on hysteresis have been written by students and associates of Childs (Collis-George, 1955; Youngs, 1958; Poulovassilis, 1962, 1970; Poulovassilis and Childs, 1971; Poulovassilis and Tzimas, 1974; and others).

Of the many contributions of the Rothamsted Laboratories to soil physics one of the most important is that of William B. Haines (1890–1962), whose series of five papers under the general title of "Studies in the Physical Properties of Soils" covered a wide range of soil physics topics. The first three of these papers published in 1925, were "I. Mechanical properties concerned in cultivation," "II. A note on the cohesion developed by capillary forces in an ideal soil," and "III. Observations of the electrical conductivity of soils." The last two of the papers, published in 1927 and 1930, were "IV. A further contribution to the theory of capillary phenomena in soil," and "V. The hysteresis effect in capillary properties, and the modes of moisture distribution associated therewith." Another paper (1950b) also dealt with hysteresis phenomena. The first of these five papers dealt with soil properties and forces associated with cultivation operations, soil cohesion, soil plasticity, and the surface friction between soil and metal and involved use of Atterberg equipment (Atterberg, 1912). A second paper (1930b) dealt with the existence of "two equilibrium series in capillary phenomena."

A. N. Puri (1898–1971) (Ramamoorthy, 1972) was at Rothamsted for a time and published numerous papers on and related to soil physics, such as papers on the hygroscopic coefficient (1925) and on the capillary tube hypothesis (1939), and a review of soil physics (1936). Many others were published with members of the Rothamsted Laboratory staff, such as one on vapor pressure and water content with E. M. Crowther and B. A. Keen (1925). Puri published a book entitled *Soils, Their Physics and Chemistry* in 1949.

Sir John Russell, second Director of the Rothamsted laboratory, was honored on his 90th birthday in an issue of *Soil Science* dedicated to him (Bear, 1962). The issue includes informative reviews of soil physics research in Great Britian and his own paper, "The rebirth of soil science in Great Britian" (E. J. Russell, 1962).

In Russia the outstanding soil physicist of the period was A. A. Rode (1896–1979). Rode worked at the Dokuchaev Soil Institute during the period 1928–1979 and wrote several books (1947, 1955, 1956). He participated with the author, S. L. Rawlins, C. E. Kellog, and T. J Marshall in preparing a chapter, "Hydrophysics of arid and irrigated soils" (W. H. Gardner, *et al.*, 1973) for an international source book on *Irrigation, Drainage and Salinity* (K. A. Kovda *et al.*, 1973). He could be regarded as the "grand old man" of Russian soil physics and, with crutches under each arm, he was an important participant in numerous meetings of the International Society of Soil Science. Alexsandr Feodorovich Lebedev's book, *Soil and Ground Waters* (1918), and his work on classification of soil water referred to earlier were important Russian contributions. Typical of many Russian contributions are a 1935 paper, "The dynamics of soil moisture," by P. I. Akopov and a 1956 paper, "Theory of equilibrium and migration of soil moisture at various degrees of wetting," by B. V. Derjagin, Madam M. K. Melnikova, and S. V. Nerpin. Numerous other less well-known Russian soil physicists have contributed to the science. Much of this work is covered in a highly comprehensive 1967 book *Physics of the Soil* (English translation, 1970) by S. V. Nerpin and A. F. Chudnovskii. An important treatise on soil science also was written by the Russian K. D. Glinka (1931)

IX. Acceleration of the Sciences Following World War II

The problem of compiling a history of soil physics is compounded with an almost overwhelming increase in soil physics research and activity of soil physicists toward the middle of the 20th century, corresponding to the marked increase in scientific research evident in the post-World War

II years. J. R. Philip, in his paper "Fifty years progress in soil physics" (1974a), estimates that 90% of all work ever done in soil physics was done in the period 1924–1974. He estimates also that "the soil physicists living today constitute rather more than 80% of those who have ever lived." Now, 12 years later, such percentages would be appreciably greater. This also is illustrated by the author's survey of numbers of soil physicists based on published papers in soil physics (W. H. Gardner, 1977) and by a cumulative curve on soil physics books written over the years, as given in a 1977 paper by Dale Swartzendruber (1925–), both of which rise steeply as the 20th century advances. Therefore, an attempt to summarize activity in the field of soil physics, including mention of all of those engaged in such work, becomes virtually impossible. A very few of the many names are included in the text with only limited biographical information. However, the listings are by no means inclusive and undoubtedly miss many important references. What is given here can only be a sampling, without critical review, of the modern work, with mention of only a few of the major themes as seen by the author. It is evident that more through historical reviews in the future must necessarily be written along more narrow subject matter lines rather than over the entire field of soil physics.

Over the years scientific symposia, conferences, and meetings have served the very useful function of keeping people abreast of current work. Large meetings, however, particularly those where concurrent sessions of competing interest are held, do not totally serve the need. A form of conference that supplements, but does not replace, the usual scientific society meeting is a midcentury product in the United States and deserving of historical note. This is the regional research program sponsored by the U.S. Federal Government cooperatively, primarily with Land-Grant Universities and Federal agencies involved in agricultural research. In 1947 the author was a member of one of the first such regional research groups, the W-9 Western Regional Technical Committee for research on drainage. Organized on a regional basis, technical committees, originally composed of no more than one or two scientists from each cooperating state, meet annually to present research reports, discuss progress, and make future plans. The existence of such groups makes it possible for each university or research unit to have, in effect, several nonresident experts in many disciplines "on" or close to its own staff, because ideas are shared in some detail each year. This type of organization has much to commend it from the point of view of avoiding duplication and keeping scientists informed beyond their own narrow specialities. Moreover, if some of the misplaced pressure for research publications by research administrators were to be removed, cooperation in publishing made possible in joint reserach efforts could go a long way toward eliminating premature and incomplete reports, thus helping to stem the publications explosion.

A dominant theme since early scientific studies began (Osborne, 1887; Oden, 1915, 1925) has been the characterization of the constituents of soil. A significant part of Keen's book *The Physical Properties of the Soil* (1931) was devoted to mechanical analyses and mechanical properties of soil. This interest continued in the postwar years with such studies as that by J. B. Page (1914–), who in 1948 reported on use of a pressure picnometer for measuring the pore space of soil. In 1950 Paul R. Day described the physical basis of the particle size analyses made using the hydrometer method much earlier developed by Bouyoucos (1927) and in 1953 obtained experimental confirmation of hydrometer theory. Day wrote the chapter on particle fractionation and particle-size analysis (1965) in the book *Methods of Soil Analysis (Black et al.*, 1965).

At mid century means for characterizing the energy state of soil water were reasonably well known, although measurement over the entire range of interest remained difficult. L. A. Richards, in his 1960 paper "Advances in soil physics," presented in the Seventh Congress of the International Society of Soil Science in Madison, WI, discusses the problem and makes a case for the use of the term "suction," with a modifier, "total," "matric," or "solute," indicating which type of forces is involved. These would be positive quantities replacing such terms as "moisture tension," "capillary potential," and "osmotic pressure." Richards refers to Theo John Marshall's (1907–) review in the book *Relations between Water and Soil* (1959), where C. G. Gurr is credited with making the original suggestion, and to his own use of suction terminology. He expresses also the hope that "some of the complicated questions ably set forth" by Bolt and Frissel (1960) and relevant to the issue will be resolved and refers to a paper by Bolt and Miller (1958) regarding calculation of the component potentials.

Earlier, Schofield, in his paper "The pF of the water in soil" (1935), introduced the use of the base 10 logarithm of the potential energy per unit mass of water, given in head units in centimeters of water, as the pF. This term gained wide use, particularly in Europe. In more recent times, however, pF terminology has been used less frequently than the terms "potential" and "suction" with appropriate modifiers and with potential being a negative quantity.

In general terms *potential* is defined as the potential energy required to move a unit quantity of water from its existing state to a state defined as zero. The total potential would be that energy required to move a unit quantity of water to a flat, pure water surface at the same elevation. Treatment of potentials varies but one common version involves recognition of five potentials: (1) gravity; (2) matric forces, involving forces of attraction between particle surfaces and water and cohesive forces in water itself; (3) osmotic forces; (4) pneumatic forces, involving pressures existing in the gas phase; and (5) overburden forces, involving the weight of substances, other than pure water, that are free to move. The latter two potentials are considered to have importance only under special circumstances. For water above the water table the matric potential is replaced by liquid pressure.

The subject of terminology is dealt with periodically by a Terminology Committee of Commission I (Soil Physics) of the International Society of Soil Science, the most current report dealing with expressions of water potential in some detail having been issued in 1974 and published as a Bulletin of the ISSS (International Society of Soil Science, 1974; see also Aslyng, 1963). The concept of water potential has had such great importance in dealing with the energy status of water and water solutions that numerous papers and critiques have been written, such as the 1961 paper by Arthur T. Corey (1919–) and W. D. Kemper (1928–), two 1961 papers by Philip F. Low (1921–), a paper by S. Iwata in 1972, and a 1985 review by Corey and Arnold Klute. Low was president of the Soil Science Society of America in 1973.

Soil suction measurements using tensiometers, limited to the range of zero to 1 bar, were discussed by S. J. Richards (1965) in a chapter in Part I of the book *Methods of Soil Analysis* (Black *et al.*, 1965). It should be noted here that A. J. Peck and R. M. Rabbidge in 1969 reported a method to extend the limited range of the tensiometer by introducing an osmotic solution in the tensiometer cup, which in this case is a semipermeable membrane. Also, it is of some interest to note that in 1947 C. G. Gurr discussed use of another type of energy measurement, freezing-point depression, in relation to the permanent wilting percentage. Stephen L. Rawlins (1971) provides a useful summary of the new methods for measuring the components of water potential. The 1972 book *Soil Water* (Nielsen, *et al.*, eds.) produced by a western United States soil water regional technical committee, deals comprehensively with all phases of soil water potential and unsaturated flow. Swelling pressure, an energy component of potential receiving less attention than most other factors, has been treated in a paper dealing with montmorillonite by B. P. Warkentin, G. H. Bolt, and R. D. Miller (1957).

Few areas of soil water physics have received as much attention as has unsaturated hydraulic conductivity or water-content-dependent diffusivity, largely because of the importance of this factor to the water-flow process. Soil water physicists—since the time of Buckingham and to some degree even earlier—have understood that water content, or the size of the water-filled channel, has a strong influence on rate of flow. Nonetheless, much of the qualitative thinking about flow has involved saturated flow concepts. Over the years numerous practical management decisions have been based on the myths that water always moves rapidly in the presence of a steep potential gradient (or water-content gradient) and that water flows rapidly into and through coarse sands and gravel. Drain lines located in wet soil above the water table, which never remove any water, are one example. Another is classification of land with fine soil overlying sands and gravels as "drouthy" when, in fact, such lands under irrigation retain water very well.

L. A. Richards, in a 1950 paper "Laws of soil moisture," clearly draws attention to a need to identify in a concise way certain observations and facts that have important applications. He proposed that two familiar soil water phenomena be referred to as laws. One of these was the Darcy equation for saturated flow and the other was what he called the outflow law. His simple statement of the latter law follows: "Outflow of free water from soil occurs only if the pressure in the soil water exceeds atmospheric pressure." He uses entry of water into underground drains as one example and water retention in fine soil above a coarse layer as another. Although, as he explains, water does not move into a coarse-textured material until the pressure comes practically to atmospheric, a more general explanation of flow phenomena that occur in pore-size-stratified materials requires examination of the water-content-dependent conductivity term in the unsaturated flow equation. Even in the presence of a steep matric potential gradient, when a wetting front arrives at a coarse layer water movement into the coarse material is slowed by the gross reduction in the cross section of the liquid flow channel. Except in the presence of a surface wetting problem, which also can have a profound influence, flow continues along surfaces and at points of particle contact. However, the flow may be reduced to a negligible rate for long periods of time while the matric potential rises in the finer soil material and larger and larger pores become filled. This flow phenomenon is a spectacular example of the influence of the unsaturated conductivity term in the flow equation. Walter H. Gardner (1917–) and J. C. Hsieh, in a 1960 time-lapse motion picture "Water Movement in Soil," described in a popular *Crops and Soils* article (Gardner, 1962, rev. 1968, 1979), demonstrates graphically most of these basic principles of unsaturated soil.

Water flow and transport has been a major part of soil physics research. At midcentury basic equations for quantification of saturated flow under diverse boundary conditions existed and the nature of the saturated hydraulic conductivity was well understood. Equations for unsaturated flow (L. A. Richards, 1931) were considered to be similar to those for saturated flow but the hydraulic conductivity, which was known to depend on water content, was not well understood. In 1946 J. C. Russell discussed movement of water in soil columns and the "theory of control sections." E. C. Childs and N. Collis-George, working with sand, had written the equation in diffusion form (1948), and at the International Soil Science Congress in Amsterdam they discussed their work on the movement of soil moisture in unsaturated soil (1950b). In 1949 Don Kirkham and C. L. Feng reported on tests of diffusion theory and laws of capillary flow and Arnold Klute (1952a, b), working on his doctoral dissertation at Cornell, used a heat-flow analogy and arrived at a diffusion solution involving a water-content-dependent unsaturated conductivity. Measurement of the diffusivity involved was treated by R. R. Bruce (1926–) and Klute (1956). In 1961 R. J. Kunze and Don Kirkham reported on capillary and self-diffusion of soil water. The diffusion approach became widely used in dealing with unsaturated flow, and, using gamma-ray measurements on water flow into a soil column, S. L. Rawlins and W. H. Gardner (1969) showed that the use of diffusion theory for describing unsaturated flow gave results similar, although not identical, to potential theory.

The diffusion approach has permitted extension of flow considerations to the vapor state. Fundamental to such considerations is the thermodynamic equation relating the humidity of the soil atmosphere to the water potential in soil. John R. Philip (1927–) and D. A. De Vries in 1957 put such flow in perspective in their classic paper "Moisture movement in porous materials under temperature gradients." The application of thermodynamics to soil moisture problems has been extensive. Among numerous papers concerned with thermodynamics of soil-water systems are one by Jolm S. Robins (1925–) in 1952 dealing with his Ph.D. research, and one by K. L. Babcock (1926–) and Roy Overstreet in 1955. P. F. Low and D. M. Anderson (1927–) in 1958 used osmotic pressure relations for determining thermodynamic properties of soil water. S. A. Taylor (Figure 19) and J. W. Cary (1931–) in 1964 reported development of linear equations for the simultaneous flow of matter and energy in a continuous soil system, and in 1969 P. H. Groenevelt and G. H. Bolt used nonequilibrium thermodynamics on soil-water systems. Munna Lal Sharma, Goro Uehara (1928–), and J. Adin Mann, Jr., in 1969 reported on the thermodynamic properties of water adsorbed on dry soil surfaces, and Groenevelt and J. Y. Parlange in 1974 dealt with the thermodynamic stability of swelling soils.

The author in his dissertation (W. H. Gardner and W. Gardner, 1951) also dealt with the flow equation, writing it in potential form and carrying out experiments to show the strong dependence of unsaturated conductivity on water content. This approach, although physically sound, is mathematically difficult and has not proved as useful in solutions to many applied problems as has been diffusion analysis, particularly in dealing with infiltration (Philip, 1954b, 1955, 1957).

In 1954 Canadians W. J. Staple and J. J. Lehane wrote on the movement of water in unsaturated flow and Australians T. J. Marshall and C. J. Gurr followed the movement of water and chlorides in relatively dry soil. W. R. Gardner in 1958 and 1959 treated the problem of flow of water in drying soil and evaporation from a water table. D. R. Nielsen (1931–), Don Kirkham, and W. R. Van Wijk (1959) dealt with measuring water stored temporarily above field moisture capacity. D. E. Elrick (1931–), writing in the first issue of the *Australian Journal of Soil Research* in 1963, discussed the unsaturated flow properties of soils. D. Zaslavsky in 1964 discussed the use of saturated and unsaturated flow equations in an unstable medium. Dale Swartzendruber discussed soil water as described by transport coefficients and functions in an *Agronomy* monograph in 1966. In 1969 H. J. Morel-Seytoux (1932–) wrote an introduction to flow of immiscible liquids in porous media and Swartzendruber wrote a chapter on the flow of water in unsaturated soil in the De Wiest book, *Flow in Porous Media* (De Wiest, 1969). E. E. Miller and A. Klute discussed mechanical forces (Part I) and J. W. Cary and S. A. Taylor treated temperature and solute effects (Part II) in a chapter on "The dynamics of soil water" in the book *Irrigation of Agricultural Lands* (R. M. Hagan *et al.*, 1967). Cary also considered the effects of the thermal regime and ambient pressures on soil drying (1967). Numerous papers have treated transport of solutes in soil water, among which are contributions by J. W. Biggar (1928–) and D. R. Nielsen on miscible displacement and leaching (1967) and by W. A. Jury (1946–), W. R. Gardner, P. G. Saffigna, and C. B. Tanner, who present a model for predicting simultaneous movement of nitrate and water in a loamy sand (1976).

Water entry into soil by rainfall or irrigation has received a significant amount of attention, with hundreds of papers being published, only a few of which are noted here. John R. Philip has written numerous papers relating to infiltration, including one on an "infiltration equation with physical significance" (1954b) and one in 1957 presenting the "Theory of infiltration: 1. The infiltration equation and its solution." In 1964 E. G. Youngs and A. J. Peck showed the nature of the moisture profile and air compression during water uptake into bounded porous bodies. This was followed by two papers by Peck (1965a, b) dealing with the same subject. P. A. C. Raats reported work on steady infiltration from line sources and furrows in 1970, and in 1973 discussed unstable wetting fronts in uniform and nonuniform soils. In 1971 Jean-Yves Parlange presented a theory for one- and two-dimensional infiltration.

Irrigation studies often have been accompanied by studies of redistribution of water following irrigation. L. A. Richards and D. C. Moore in 1952 considered the effect of capillary conductivity and depth of wetting on water retention. In 1969 Eshel Bresler, W. D. Kemper (1928–), and R. J. Hanks (1927–) reported a study of infiltration, redistribution, and subsequent evaporation from soil as affected by wetting rate and hysteresis. R. D. Jackson and F. D. Whisler (1970) showed equations for approximating vertical non-steady-state drainage of soil columns, and in 1971 K. F. Kastanek developed a numerical simulation technique for vertical drainage from a soil column. Two papers by E. G. Youngs (1957, 1958) dealt with redistribution in profile water content after irrigation. E. C. Childs (1964) showed the nature of the ultimate moisture profile during infiltration into a uniform soil. W. R. Gardner, D. Hillel, and Y. Benyamini (1970) show families of curves representing profile water change during redistribution, illustrating the dynamic nature of water retention. Many modern studies of redistribution illustrate the falacy of long-standing misconception by numerous nonspecialists that a unique field capacity exists. Perspective on field capacity as a dynamic concept is given in Chapter 10 of the text *Soil Physics* (4th ed., L. D. Baver, W. H. Gardner, and W. R. Gardner, 1972).

Retardation of infiltration into layered soil was shown by David E. Miller (1926–) and W. H. Gardner in 1962. Graphic demonstration of infiltration and advance of wetting fronts into layered soils was shown in the Gardner and Hsieh 1960 time-lapse motion picture mentioned earlier. D. Hillel (1930–) and W. R. Gardner dealt with infiltration into crust-topped profiles under steady (1969) and transient (1970a) conditions and measured the unsaturated conductivity and diffusivity by infiltration through an impeding layer (1970b). The general subject of infiltration into layered soils was treated by R. Russell Bruce and F. O. Whisler (1973) in a chapter in the book *Physical Aspects of Soil Water and Salts in Ecosystems* (Hadas *et al.*, 1973).

Another refinement to flow theory receiving considerable attention is the recognition that matric potential is not a single-valued function of the soil water content but depends on whether the soil has been wet up or dried down to the state in question. As mentioned earlier Childs and others had recognized the hysteresis problem. E. E. Miller and R. D. Miller were early contributors in this area with a 1956 paper "Physical theory for capillary flow phenomena." Philip (1964), Topp (1937–) and Miller (1966), Topp (1971), G. Vachaud (1939–), and J. L. Thony (1971), and others, in addition to those associated with Childs, have contributed to these studies.

Although the general nature of unsaturated conductivity, hydraulic conductivity, or capillary conductivity as it was called earlier, had been known qualitatively for some time, its quantitative description awaited methods to measure the flux and the moving force term in the unsaturated flow equation. In his 1931 paper L. A. Richards used steady-state flow and tensiometers to make measurements in short columns over the limited tensiometer range. In 1953 S. J. Richards and L. V. Weeks obtained capillary conductivity values from moisture yield and tension measurements in soil columns, again only over the limited tensiometer range. However, with the development of a pressure-membrane apparatus (L. A. Richards, 1941) it became possible to make measurements over the entire range of interest. This was done in 1956 by Wilford R. Gardner, a student of Kirkham. Gardner was elected to the U.S. National Academy of Sciences in 1983. He measured the outflow from a pressure-membrane apparatus as pressure was increased in small increments. This method has been verified and refined by E. E. Miller and D. E. Elrick (1958), Rijtema (1959), Kunze and Kirkham (1962), and Peck (1966).

An infiltration method for measuring unsaturated conductivity was described in 1964 by E. G. Youngs, and in the following year D. Zaslavsky and I. Ravina reported a method based on moisture movement. An instantaneous profile method was reported by K. K. Watson (1928–) in 1966. Arnold Klute has worked extensively in this field and has written a comprehensive review (Klute, 1972). Modern radiation techniques for measuring water content nondestructively in laboratory columns and in field situations have made it possible to make measurements in long columns and to improve on the laboratory methods for determining unsaturated conductivity.

Another approach, determination of unsaturated conductivity from moisture-retention data, has received considerable attention recently in association with modeling studies of water flow and of overall soil-plant-atmosphere systems. Early efforts along this line were made by E. C. Childs and N. Collis-George (1950a), in 1958 by T. J. Marshall, and in 1959 and 1960 by R. J. Millington (1926–) and J. P. Quirk (1924–). In 1960 D. R. Nielsen, Don Kirkham and E. R. Perrier compared measured and calculated values. Larry Boersma (1930–) covered field methods for measuring hydraulic conductivity in the "methods book" (Black *et al.*, 1965). Don Nielsen succeeded the author as president of the Soil Science Society of America in 1984, Larry Boersma became President Elect in 1985, and another soil physicist, Anson R. Bertrand (1923–), who has spent much of his career in soil science administration, was president in 1974. Gardner also served as Editor-in-Chief of the Soil Science Society of America Proceedings from 1966–1969.

Comparisons of pre-1971 efforts with each other and with measured values using other techniques were made by R. E. Green and J. C. Corey (1971). In 1972 R. Russell Bruce (1926–) calculated hydraulic conductivity from water-retention relations. R. D. Jackson (1929–), a student of W. Doral Kemper (1928–), analyzed some of the data and proposed modifications in the equations (Jackson, 1964). In 1974 Gaylon S. Campbell, a student of W. H. Gardner, reviewed some of these approaches and proposed an equation similar to some used by Hillel and Gardner (1969) and given by D. Hillel in his book (1971). He shows data from some of the papers he discusses along with recalculation using his equation.

Considerable progress has been made on measurement of matric potential in the postwar era. Recognition that vapor pressure in the soil atmosphere could be converted to total suction (or matric plus osmotic potential) led L. A. Richards and G. Ogata in 1958 and J. L. Monteith and P. C. Owen in England in the same year to use a special psychrometer to measure relative humidity in the soil atmosphere. The first measurements of this type were limited by the necessity for adding a small amount of water to the thermocouple for the necessary evaporation. However, D. C. Spanner in England in 1951 had used the Peltier effect to condense water on a thermocouple. Use of this adaptation made in situ measurements possible (Korven and Taylor, 1959). Initially the method was limited by a requirement that temperature be controlled to the order of 0.001°C, but this limitation was eased when in 1967 Stephen L. Rawlins (1932–), a student of W. H. Gardner, and F. N. Dalton demonstrated that reducing fluctuations in temperature down to this order of magnitude over the time of measurement was all that was required. This made it necessary in the laboratory only to provide a large thermal mass surrounding the measuring unit. Also, measurements could be made in field situations at depths where temperature change over time was within the necessary limits.

Laboratory measurements were the earliest to be developed and to facilitate these a sample changer with multiple chambers for holding small samples was developed (Campbell *et al.*, 1966). These are in wide use in obtaining data for sorption and desorption curves, with corrections being made for the osmotic potential where it is large enough that it cannot be neglected, as well as in handling small samples of many kinds. Thermocouple psychrometry was extended to measurement of leaf water potential (the sum of osmotic and matric potentials), but with some reduction in accuracy because of difficulties encountered in holding uniform temperatures. Extension of the method to biological materials has led to wide use in other areas of science. Use in plant pathology was particularly important, where it became possible to study pathology problems in environments where moisture stress could be precisely known [R. J. Cook (1932–) and R. I. Papendick (1931–), 1970; Papendick and Campbell, 1974]. Measurements were made in trees (Wiebe *et al.*, 1970) and new insights into seedling replant survival became possible. Measurements of osmotic potentials in blood for the medical profession and in food processing were a few of the many applications being made of the thermocouple psychrometer.

Although the psychrometer method requires considerable care in use, it has extended the range of water potential measurement in soil to well below the tensiometer range and, in fact, to materials with water potentials well below the range of interest in soil. Hundreds of studies have been made in the development of the thermocouple psychrometer (and a related dew-point hygrometer) only a few of which are cited here. Rawlins' paper on the theory of the method (1966), and a subsequent theory paper by A. J. Peck (1968), helped materially in the development.

The psychrometer method was later extended to make it possible to separate the matric from the osmotic potential (Ingvalson *et al.*, 1970). This involved obtaining the osmotic component from measurement of the electrical resistance of an accompanying finely porous ceramic unit that remained saturated. Such units had been described in 1959 by W. Doral Kemper (1928–). Kemper became interested in soils through T. L. Martin at Brigham Young University and received his Ph.D. with J. Fulton Lutz at North Carolina.

As in other soil situations, temperature is an important factor in psychrometer measurements; however, precise temperature control is of lesser importance. H. R. Haise (1914–) and O. J. Kelley (1914–) in 1950 had been concerned about large diurnal fluctuations in tensiometer readings and were able to show that the fluctuations were caused primarily by temperature gradients from tensiometer cup to soil and that the temperature change at the cup largely resulted from heat conduction through the tensiometer body itself. Robert Gardner (1888–1977) in 1955 showed experimentally using tensiometers that "soil moisture tension" decreased only about 8 cm of water per Celsius degree. Sterling A. Taylor (1958), S. A. Taylor and G. S. Stewart (1960), G. E. Wilkinson and Arnold Klute (1962), Klute and L. A. Richards (1962), J. W. Kijne and Taylor (1964), and Gaylon S. Campbell and W. H. Gardner (1971) showed generally small temperature effects, but some in both directions. These usually would be negligible at field temperatures except, possibly, in very dry soil.

That temperature is an important property of soil from many points of view is evident from the extensive literature on the subject. A variety of papers typical of this literature are such as that by C. G. Gurr, T. J. Marshall, and J. T. Hutton describing water flow resulting from a temperature gradient (1952); the 1952 treatment of soil temperature and plant growth by S. J. Richards, R. M. Hagan (1916–), and T. M. McCalla in the B. T. Shaw book, *Soil Physical Properties and Plant Growth*; the 1957 paper of M. E. Bloodworth (1920–) and J. B. Page on the use of thermistors for measurement of soil moisture and temperature; the 1962 Craig L. Wiegand and Sterling A. Taylor paper on temperature depression and distribution in a drying soil column; the 1963 chapter on the thermal properties of soil by D. A. De Vries in the W. R. van Wijk book, *Physics of Plant Environment*; the 1965 treatment of water flux in moist soil comparing thermal versus flux gradients by J. W. Cary; and a 1975 paper by G. R. Mehuys. L. H. Stolzy (1920–), and J. Letey (1933–) on the temperature distribution in soil under stones submitted to a diurnal heat wave.

Although water content is explicitly or implicitly involved in nearly all practical and theoretical problems in soil science, its measurement has been difficult. It has been necessary either to take samples laboriously and weigh and dry them to compute the water content, or to measure some other property of moist soil or of a buried device whose action depends on water content and to infer the water content. In addition to experimental errors associated with handling soil samples, sampling involves uncertain errors related to spatial variability of soil. Moreover, where moisture-sensitive devices are used, such as a porous block with electrodes or a tensiometer cup, stability and problems related to multiple-valued relations dependent on wetting history often lead to measurement uncertainties.

A number of moisture-sensitive devices have been used, such as the Bouyoucos electrical block mentioned earlier or similar blocks with some other property measured, such as electrical capacitance (Anderson, 1943). Byron T. Shaw and L. D. Baver in 1939 and 1940, and others since, added a small amount of heat in a block and measured heat conductivity as an index of moisture content. Joel E. Fletcher (1911–) (1940), and in 1948 Thorne and Russell, measured the dielectric properties of a block as an index of water content. Marlowe D. Thorne (1918–) was a student of Russell's at Iowa State and after the war at Cornell, where he received his Ph.D. He later became a department head with the Pineapple Research Institute at the University of Hawaii, was project leader in the eastern United States for the USDA–ARS program, chaired departments at Oklahoma State and Illinois, and spent 2 years on a technical assignment in India. He was the President of the American Society of Agronomy in 1977.

Application of nuclear radiation methods to measurement of soil properties, beginning at midcentury, appreciably extended capabilities for making water-content measurements and improved their reliability, at least under some circumstances.

Rapid developments in nuclear science following war use of nuclear energy led two groups, one involved with aeronautical research and the other in soil science and working independently, to apply neutron-scattering and thermalization to measurement of soil water content (Belcher *et al.*, 1950; W. R. Gardner and Kirkham, 1952). This measurement was the basis for W. R. Gardner's Master's thesis under Kirkham at Iowa State University. The method makes possible highly accurate measurements of overall water content below about the top 0.15 m and is nondestructive after installation of a metal access tube. Although spatial resolution is poor, this method does effectively eliminate confusion of changes in time with changes from point to point in the field that confound gravimetric sampling techniques. Application of this new method of water-content measurement has led to a vast literature both in soils research and in applied management situations. Background material on the technique and some of the pertinent literature are given in a chapter on water content in the book *Methods of Soil Analyses*, Part 1 (W. H. Gardner, 1965; updated version in a new edition of the "methods book," in press, and to be published in early 1986). Useful practical information on the method is found in a 1958 publication by J. W. Holmes (1921–) and K. G. Turner. The method also is described in an irrigation monograph (Holmes *et al.*, 1967). M. Visralingham and J. D. Tandy (1972) have reviewed the literature on the neutron method.

Use of gamma rays to measure soil bulk density and water content, another outgrowth of postwar studies in radiation physics, has greatly improved research in soil water. R. K. Bernhard and M. Chasek in 1953 and J. A. Vomicil in 1954 measured soil bulk density, and F. M. Ashton (1922–) in Hawaii in 1956 reported use of gamma rays to follow water content changes in soil growing sugar cane. This method was further developed by C. H. M. van Bavel (1921–), N. Underwood, and S. R. Ragar (1957), van Bavel (1959), Gurr (1962), H. Ferguson and W. H. Gardner (1962), Davidson *et al.* (1963), and numerous investigators since.

Extension of gamma-ray methods for concurrent measurement of both mineral bulk density and water content was achieved independently and reported in 1967 at an Istanbul symposium "Isotope and Radiation Techniques in Soil Physics and Irrigation Studies," sponsored by the International Atomic Energy Agency, by B. D. Soane (1967) of England and by the author and C. Calissendorff (1967). Soane's measurements were made without collimation of gamma beam and were applied to studies in the field, whereas Gardner and Calissendorff used highly collimated beams for high spatial resolution measurements in the laboratory (W. H. Gardner *et al.*, 1972). These methods have been used extensively and applied particularly to water-flow situations in swelling and shrinking soils (Soane, 1968; Nofziger and Swartzendruber, 1974).

From the earliest years of the science soil structure has received serious attention from soil scientists generally as well as from soil physicists. Understanding the development of soil structure and the use of both natural and synthetic soil-conditioning materials to alter soil structure involves the consideration of several areas of soil science, including physics, mineralogy, physical chemistry, biochemistry, and microbiology. Basic studies on the interaction between clays and organic compounds were reported in England by D. J. Greenland (1965). Aggregate stability by Kemper and bulk density by Blake are the subjects of chapters in the book on *Methods of Soil Analysis* (Black *et al.*, 1965). In 1960 James M. Davidson (1934–) and D. D. Evans (1920–) described a turbidimeter technique for measuring the stability of soil aggregates in water–glycerol mixtures and W. C. Moldenhauer (1923–) and Kemper (1964) evaluated the interdependence of water drop energy and clod size in infiltration and clod stability. D. D. Evans and Don Kirkham (1949) used air permeability as a measure of soil structural properties; in 1950 J. P. Quirk (1924–) in Australia reported on measurements of the stability of soil microaggregates in water; in 1953 Vernon C. Jamison (1907–1968) in the United States dealt with changes in air–water relationships caused by soil structural improvement; and A. J. Low (1954) in England reported studies of soil structure in the field and laboratory. J. P. Quirk and R. K. Schofield (1955) noted the effect of electrolyte concentration on the structure of a soil and its consequent permeability.

Johan Bouma in The Netherlands, whose interests include soil classification and survey as well as soil physics, has considered expansion of soil survey interpretations through greater use of physical methods, particularly in dealing with drainage conditions (1973). In the same year, with J. L. Anderson, he has discussed the relationships between soil structural characteristics and hydraulic conductivity in the monograph "Field Water Regime" (Bruce *et al.*, 1973). W. D. Kemper and

E. J. Koch (1966) reported on a comprehensive survey of aggregate stability of soils collected from the western United States and Canada. In 1958 a Russian, P. V. Vershinin, dealt with soil structure in a book *The Background of Soil Structure*.

Dealing with an important moisture-dependent property of soil, J. S. Vomocil, E. R. Fountaine, and R. J. Reginato (1958) measured the effect of moisture content on the tensile strength of glass bead systems, and in 1961 Vomocil, with L. J. Waldron and W. J. Chancellor, measured soil tensile strength by means of centrifugation. J. M. Cary and D. D. Evans (1974) reported on an extensive regional research study of soil crusts, and R. C. Reeve (1920–) in 1965 wrote the chapter on modulus of rupture, first described by L. A. Richards (1953), in the book *Methods of Soil Analysis* (Black *et al.*, 1965)

An important postwar development has been the discovery of synthetic soil conditioners. Soil porosity as affected by aggregation and the influence of cultivation and soil additives, such as manures, has been studied from the earliest of times. However, the ability to alter such structural features of soil significantly for any appreciable length of time did not exist until the discovery of synthetic soil conditioners. In the waning years of the war a Canadian microbiologist–biochemist, J. H. Quastel (1899–), working in England, discovered that extremely small quantities (as little as 0.05% by weight) of certain long-life synthetic polymers had a pronounced stabilizing effect on soil (Quastel, 1952). These polymers resemble natural soil-stabilizing materials, polysaccharides and polyuronides that come from decomposing organic matter. Quastel communicated this information to his friend, Charles Thomas, of the Monsanto Chemical Co. in St. Louis, Missouri, who became sufficiently interested to inaugurate a research program. This research led to the announcement in a 1951 symposium ("Improvement of Soil Structure," Philadelphia meetings of the American Association for the Advancement of Science with six papers published in Volume 73, *Soil Science*, 1952: Quastel; Hendrick and Mowry; Allison; Martin, Taylor, Engibous, and Burnett; Weeks and Colter; and Ruehrwein and Ward) of the discovery of the aggregate-stabilizing properties of synthetic polymers, vinyl acetate maleic anhydride (VAMA), and hydrolyzed polyacrylonitrile (HPAN).

Because soil porosity or soil structure is at the heart of so many problems involving soil, anything that influences structure is of immediate interest to a broad spectrum of people who work with soil. This spectrum extends from soil experts, farmers, and nurserymen, who are known to expend great effort modifying soil for particular purposes, through conservationists and home gardeners. A 1953 paper describing the effect of soil aggregating chemicals on soils was written by G. S. Taylor (1920–) and W. P. Martin (1912–). Possible influence on salt-affected soils was of particular interest (Allison and Moore, 1956; Carr and Greenland, 1975; Peters *et al.*, 1953). Newspaper headlines and feature articles began to appear throughout the United States and elsewhere and a new industry was born. In the early 1950s the new synthetic soil conditioners became available experimentally and research soil scientists everywhere began experiments. These showed that the material indeed would stabilize a desirable soil structure—as well as an undesirable one with improper soil preparation—and promote seedling emergence and plant growth in soils where mechanical properties and aeration were limiting.

Although Monsanto had intended to move more slowly in marketing the material, pressure developed from prospective users and from other chemical companies, which already marketed HPAN as a drilling mud for the petroleum industry and now were offering it as a soil conditioner. As a consequence Monsanto put VAMA on the market under the name Krilium and HPAN under the name Bondite. Although costs were prohibitive for large-scale application on farmlands, its use soon became extensive in home gardens, greenhouses, and nurseries and on high-value "crops" such as golf courses and playing fields, where soil structure indeed was a limiting factor.

Hundreds of scientific papers were written—only the six original symposium papers and a few others are cited here—and numerous symposia and conferences have been held. The author presented a review of the synthetic soil conditioner development in a Ghent, Belgium symposium in 1971 (W. H. Gardner, 1972). The excitement over synthetic soil conditioners died down after a few years, primarily because of the high cost of materials and their application, but additionally because successful use requires careful manipulation of the soil following application to form the structure to be stabilized, and many users were unable or unwilling to go to this trouble. Most expected to be able to spread the materials on the surface and come back the next day to find a perfect seed bed.

Since the introduction and decline in use of VAMA and HPAN, the search has continued for other effective, but more economical, materials with some success, although economic considerations have not yet favored widespread use. A symposium "Experimental Methods and Uses of Soil Conditioners," sponsored by the Soil Science Society of America and the Committee on Soil Conditioners of the International Society of Soil Science in 1975 (Moldenhauer *et al.*, 1975), has covered more recent developments and use of soil-conditioning materials. A few of the papers are listed here to provide an idea of the subject matter covered. Of particular interrest is the first paper of this symposium, by M. De Boodt, entitled "Use of soil conditioners around the world" (De Boodt, 1975). Mechanisms of stabilization were treated by N. Schamp, J. Huylebroeck, and M. Sadones (1975); Carr and Grenland (1975) described extensive favorable results from use of poly vinyl acetate and poly vinyl alcohol on

Figure 19. Taylor, Sterling A. (1918–1967). From *Soil Water*, dedicated to Dr. Taylor and published by American Society of Agronomy, 1972.

sodic soils; Armbrust and Lyles (1975) showed promising results in use of soil stabilizers to control wind erosion; and Letey (1975) and De Bano (1975) discussed the use of surfactants on water-repellent soils and their effect on water flow and retention.

Emerson (1965) and De Boodt (1972) reviewed the field of soil conditioners. An extensive treatment of this subject is found in the book *Modification of Soil Structure*, edited by W. W. Emerson, R. D. Bond, and A. R. Dexter (1978), reporting a symposium on the subject held under the auspices of the International Society of Soil Science, Commission 1 (Soil Physics) in Adelaide, Australia, August 1976. The introductory paper of this symposium, "Some physico-chemical aspects of soil structural stability—a review," presented by J. P. Quirk (1978), well illustrates the complicated nature of the subject matter; likewise, the list of papers, involving nearly 100 authors and extending from basic considerations to field applications, shows the breadth of coverage. Surfactants and water-repellent soils were treated in detail in a 1969 symposium (De Bano and Letey, 1969)

Sterling A. Taylor, whose career began and ended in the postwar years, is deserving special note. Taylor did his undergraduate work at Utah and his postgraduate work under M. B. Russell at Cornell. After he received his Ph.D. in 1949 he returned to Utah, where he engaged in reserach along several lines in soil physics. He is particularly noted for his early work on oxygen diffusion (1950), the activity of water (1958), the application of nonequilibrium thermodynamics to water flow (Taylor and Stewart, 1960; Taylor and Cary, 1964), and plant growth conditions in soil (1952; Taylor *et al.*, 1961), publishing numerous papers with his students, who are active in the field today. Taylor lived an extremely active life in his relatively short professional career, having had major influence on young people as a Boy Scout leader as well as in the university. He died of cancer in 1967, only weeks before he was scheduled to present an important paper at an international symposium in Prague. A textbook *Physical Edaphology* was completed by Gaylon L. Ashcroft and published after Taylor's death (S. A. Taylor and Ashcroft, 1972).

Because the objective of a considerable part of soil physics research is associated with plant growth, numerous studies have been made of root growth, root water uptake, and movement of water through the plant and into the atmosphere, or through what often is referred to as the soil-plant-atmosphere continuum. C. B. Tanner (1920-) and his many students have been much involved with evapotranspiration, plant-water relations, and the microclimate of plants and in measurements of the water status of plants, such as psychrometry applied to plant leaves (Tanner, 1967, 1968; Tanner and Elrick, 1958). He was early influenced by Thomas L. Martin and worked with E. E. Miller and M. L. Jackson for his Ph.D. at Wisconsin. He was elected to the U.S. National Academy of Sciences.

C. H. M. van Bavel has made evapotranspiration estimates as criteria for determining time of irrigation (1952a) and has made field measurements of water uptake by roots. He also has worked with gaseous diffusion in porous media (1952b). He was educated in The Netherlands and at Iowa State University with Don Kirkham.

In 1968 E. R. Lemon (1921-), a student of M. B. Russell's before he went to Michigan State University for doctoral work on oxygen diffusion in soil with A. E. Erickson (1919-), dealt extensively with the energy and water balance of plant communities. Work in this area, basic to management of water in the field, has depended heavily on development of energy concepts that can be applied to both soil and plant and to the development of water-flow processes. Lemon also worked extensively on soil aeration and plant growth (Lemon, 1962; Lemon and Wiegand, 1962). W. A. Raney (1920-) in 1949 reported work using oxygen diffusion as a measure of soil aeration with implications to plant growth. The field of soil aeration and plant growth was reviewed by Albert R. Grable in 1966 in *Agronomy Monographs*.

W. R. Gardner has contributed to soil-plant-atmosphere field through studies of the dynamic aspects of water availability to plants (1960). In 1962 O. T. Denmead and R. H. Shaw added to the understanding of plant water availability with studies of how availability is influenced by soil moisture and meteorological conditions. In 1965 J. R. Cowan dealt with transport of water through the soil-plant-atmosphere system. An interesting related observation of flow in this continuum was made some years earlier in 1953 when Edward L. Breazeal and W. T. McGeorge showed that water could move through the continuum in the reverse direction, adding water to the soil from other parts of the system. They tied small bags of dry soil around the stem of a tomato plant so that adventitious roots would grow into and, under the right conditions, transmit water into the soil. They had earlier studied vapor movement in soil (1951). A short biography of Frank Breazeale written by T. F. Buehrer was published in *Soil Science* in 1951.

Plant environment and efficient water use was the subject of a special symposium sponsored by the Soil Science Society of America and American Society of Agronomy at Iowa State University in 1965. The proceedings, edited by W. H. Pierre, Don Kirkham, and John Pesek, was published in 1966 and contained many papers that greatly expanded knowledge of soil-plant relations. The numerous models for water transport in the soil-plant-atmosphere system were reviewed by Moltz in 1981. J. Fulton Lutz (1952) reviewed the influence of mechanical impedence on plant growth in the book *Soil Physical Conditions and Plant Growth* (B. T. Shaw, 1952). Howard M. Taylor (1924–) and Herbert R. Gardner have worked extensively on penetration of roots into soil as related to soil strength (H. M. Taylor and Gardner, 1963; H. M. Taylor, 1971). Mechanical resistance to growth of roots was treated in 1967 by K. P. Barley and E. L. Greacen in an issue of *Advances in Agronomy*. R. E. Danielson in 1971 looked at nutrient supply and uptake as influenced by soil physical conditions.

Early efforts at irrigation scheduling had depended on following the water content of the rooting zone soil to determine when water was limiting. Tensiometers were used to follow the water status, but their limited range was some disadvantage. Numerous studies were made on use of tensiometers, particularly for such high-value crops as citrus. Typical of these studies is work reported by S. J. Richards and A. W. Marsh (1961). As an alternative to use of water content or energy status measurements for irrigation scheduling, extensive work began in the 1950s to determine soil water depletion by measuring evapotranspiration. In The Netherlands W. R. van Wijk and D. A. De Vries (1954) wrote a treatise on evapotranspiration. In 1962 W. R. Gardner and D. Hillel related external evaporative conditions to soil drying, and J. W. Cary (1967) considered the influence of thermal regime and ambient pressures on soil drying. H. R. Haise and Robrt M. Hagan wrote a chapter on "Soil, plant, and evaporative measurements as criteria for scheduling irrigation" in the monograph *Irrigation of Agricultural Lands* (R. M. Hagan *et al.*, 1967). H. R. Gardner in 1969 reported on the relation of water application to evaporation and storage of soil water. An evaluation of several methods for estimating evapotranspiration under semiarid conditions was presented in 1973 by R. J. Hanks, H. S. Jacobs, H. E. Schimmelpfennig, and M. Nimah. The subject of water content and status of water in soil has been dealt with by Rawlins (1976) in Kozlowski's *Water Deficits and Plant Growth* (Kozlowski 1968, 1976). A number of women have entered the soil physics field. Two women plant physiologists, Elizabeth

Lee (Betty) Klepper (1936–), with the ARS at Pendleton, Oregon, and Mary Beth Kirkham, daughter of Don Kirkham, now at the Evapotranspiration Laboratory at Kansas State Unviersity, Manhattan, have contributed appreciably to the understanding of plant-soil-water relations.

An application of soil water physics with significant current application and important implications to future crop production and irrigation efficiency is the modern development of trickle irrigation. Many studies have been conducted on flow from point and line sources and on the influence of such irrigation on the growth of root systems. Considerable research in this new and important area has been done by Israeli research groups. The background and present status of trickle irrigation are covered in a 1977 review by Eshel Bresler, Institute of Soils and Water in the Volcani Center at Bet Dagan, Israel, which was published in *Advances in Agronomy*. Trickle systems provide opportunities for water application at a wide range of rates and frequencies. A case for high-frequency irrigation to optimize rooting zone water content has been made by S. L. Rawlins (1973) and by Rawlins and P. A. C. Raats (1975).

Created in the postwar years, the Division of Environmental Mechanics of the Commonwealth Science and Industry Research Organization, CSIRO, in Australia has contributed significantly to soil physics. This division had its beginnings in 1951 when John Philip was appointed to investigate the hydraulics of irrigation in the Regional Pastoral Laboratory, which later became the Agricultural Physics Section of the Division of Plant Industry and, ultimately in 1971, became the nucleus of the Division of Environmental Mechanics. The task of this division was to "contribute through the development and use of physical and mathematical techniques, to our understanding of the basic mechanisms of transfer processes in the natural environment where plants, animals and people live." John Philip has taken a strong interest in the development of those agricultural sciences which involve water and has made major contributions to the solution of equations for flow of water in unsaturated media under a wide variety of physical situations. He has given comprehensive coverage to the development of infiltration theory.

Numerous Australians have been interested and involved in soil physics research, a few of whom have been mentioned already. The list includes such people as N. Collis-George, O. T. Denmead, C. G. Gurr, J. W. Homes (1921–), T. J. Marshall, R. J. Millington (1926–), Adrian Peck, J. P. Quirk, D. E. Smiles (1936–), K. K. Watson (1928–), I. White, and many others.

Many symposia on soil physics topics, both on society programs and in separate meetings and conferences, have been held around the world in recent years. Publications resulting from such symposia constitute a rich source of material synthesizing what is known about the physics of soils.

The symposia are too numerous to discuss individually, but several appear in the references cited. Two are worthy of special mention as representative of the many: first is the symposium "Water in the Unsaturated Zone," held in Wageningen, The Netherlands, and known commonly as the Wageningen symposium." The proceedings of this symposium, edited by P. E. Rijtema and H. J. Wassink (1969), continue to be a source of important information on soil water physics. The second symposium of note is "Soil Water Physics and Technology," held in Rehovot, Israel in 1971, and sponsored by the International Society of Soil Science, Commissions I (Soil Physics) and IV (Soil Technology), organized by the Israeli Soil Science Society. The 1973 publication coming out of this symposium, *Physical Aspects of Soil Water and Salts in Ecosystems*, edited by A. Hadas, D. Swartzendruber, P. E. Rijtema, M. Fuchs, and B. Yaron, contains papers by soil physicists from all over the world. The list of author names and topics could well be included here as representing soil physics in the decade of the 1960s.

X. The 1950s and Beyond

There is a great need for more interpretation and synthesis of scientific discovery. More "desk scientists" are needed who can review what has been observed, or can be derived from particular observations, and then synthesize and report on the state of the art. Computer modeling is an adjunct to this process, but not the center of it. Such endeavors might well identify gaps in knowledge requiring further experimentation or analyses as well as new directions for research.

Paradoxically, in recent years agricultural research in the United States has outdone itself to the extent that serious overproduction exists, whereas in some other parts of the world people are starving because of inability to use available land successfully or shortages of productive soil. Consequently, it is difficult to generalize on immediate need for further soils research, except on the basis of perfecting knowledge of soil and soil processes. Nonetheless, assuming that no other population checks become operative, existing soils inevitably will become inadequate to satisfy the needs of burgeoning populations. However, protecting soil requires immediate attention with respect to the role it plays in the storage and purification of water supplies. Water quality and supply currently is critical in numerous places. As land is used more and more as a reservoir for waste products of various kinds, protection of soil and water will become increasingly critical.

Soil is essential to the storage and gradual release of water to streams to provide year-round supplies of water. Soil in appropriate physical condition also acts as a filter for eroded soil particles and is the medium in which biological processes affecting water purity can take place. Soil, as the medium in which most plants are grown, must be recognized as the most critical of all natural resources because of this and because of its influence on water supply.

Soil erosion, creator of soil in vast delta planes but destroyer of large areas of soil that currently are productive, remains as one of the serious soil problems of the day. A modern development deserving continuing attention is the concept of minimum or zero tillage to reduce susceptibility to erosion of soils kept in cultivated crops. A United States National Conservation Conference "Tillage, Soil Environment and Root Growth" (Allmaras *et al.*, 1973) has reviewed tillage problems and related research and H. Kuipers (1970) has provided historical background of the zero-tillage concept.

It is evident that, on the whole, there are huge untapped soil and water resources yet available on the earth if the water can be brought to the soil. At the same time, there are promising leads to vast energy resources that may make this possible. However, such achievement would be attended by a myriad of related problems, some of which have global significance and importance. Synthesis required to identify what is known and what yet must be learned about such problems demands the efforts of the world's best minds. Innumerable problems of lesser scope exist where synthesis is required to point the way.

Much of the physical information needed to deal with many existing and expected soil resource problems already is available in the literature of soil physics. However, soil is a highly complicated colloidal material and it manifests itself in a great variety of ways in the surface mantle of the earth. The physical processes of flow of gases, heat, and water and their interaction with plant and microbiological life are highly complex. Even a qualitative description of some of the processes is difficult and quantitative description often requires higher mathematics not generally understood by all people who need to know. Oftentimes the guidance of a simple physical principle is sufficient, but where it is not, recognition of the nature of a problem can lead to a solution involving people with the needed knowledge.

The author's experience with a time-lapse motion picture "Water Movement in Soil" (with J. C. Hsieh), referred to earlier, shows one way to demonstrate important ideas in a useful and simple way. Quantitative expression of these things, which requires basic physics, physical chemistry, and mathematics well beyond the ability of many people, may not always be needed. Other movies and demonstrations of difficult principles involving soil are available, but more are needed to put into practice what already is known about soil, its use, and its conservation.

Soil scientists must anticipate problems that may be associated with developing vast new land areas. This might be important if nuclear fusion should make abundant an inexpensive power that would permit production of clean water from the oceans. These are many-faceted problems that could occupy the efforts of numerous scientific specialists, including soil physicists.

Soil physicists are not likely soon to run out of things to study. However, one of the subjects in which further effort is warranted involves description of physical properties and processes on a broad area basis to facilitate appropriate management. Considerable progress already has been made on spatial variability and the statistics involved (Warrick *et al.*, 1972; Nielsen *et al.*, 1973; Biggar and Nielsen, 1976; Philip, 1980), but much has yet to be done.

The future of soil physics was discussed in an evening session at the SSSA annual meetings in Washington, D.C. in August 1983 (unpublished). Introductory presentations were made by W. R. Gardner and S. L. Rawlins, after which numerous topics relating to present and future work in soil physics were discussed. There seemed to be a consensus that soil physicists had greatly expanded their perspectives to the point where the suitability of the title "soil physics" even might be questioned. It was noted that soil physicists have worked extensively with plant physiologists and pathologists, with entomologists, and with civil and agricultural engineers, and with environmental scientists having varying interests. Use of computer modeling and its applications to applied problems received considerable attention—also its use in guiding experimental and theoretical research. No consensus was reached regarding what research was most needed, but many people felt that a great deal more was known today than was applied to practical problems—further justification for more attention being given to synthesis of soil physics knowledge.

The author has been very much impressed by his experience in writing this history. At times he has been so totally immersed in his reading and study that even his dreams have related. It is a common experience to go to the libarary stacks for some small detail and to return with sufficient information for an entire chapter. Lack of time, journal space, and particularly the lack of ability to read and to evaluate, have been frustrating limitations in trying to present a complete history.

It is ironical that too much information is available on the one hand, and too little on the other. The lack is on the human side, and this aspect of the history has great interest to the author. An even treatment of the human history is impossible, but a small attempt nevertheless has been made with the hope that this may stimulate others to share their impressions of the personalities and backgrounds of the people who have made our science.

The author ends his history with a small selection of impressions of a type that have almost haunted him during the preparation. A few of the many memories are of young Ren and Sterling Richards in white tennis clothes with tennis rackets in the hallway outside of the office of the author's father; the mystified look on the face of a Dean (shortly to become the director of research for the newly formed Soil Conservation Service and later a state Govenor) as a very young Don Kirkham showed him experimental proof of the unexpected (to the Dean, at least) nature of streamlines in a drainage situation, and the youthful exuberance of Kirkham, not yet old enough in his appearance to have a Ph.D. and be talking to a Dean that way; and an "elderly" Don Kirkham today who appears anything but "elderly" as he takes an interest in anything and everything, such as walking over Europe with a knapsack on his back, unmindful of the many troubles that might beset him. Of the same Kirkham's relief when, having telegraphed to airports and railroad stations from Prague to Istanbul trying to locate them, he finally saw Wilf and Walt Gardner come into an Istanbul hotel a day late because of transportation problems occasioned by the 7-day Israel war. Of Richard Bradfield in Chicago as a kindly gentleman listening to a naive, young Bachelor of Physics explain to him, stupidly, that practically all of the elements in soil might be identified usefully by spectrographic analyses, thus eliminating wet chemistry—and later to experience Bradfield's lectures in a soil physics course wherein Bradfield became so wrapped up and enthusiastic on his subject that his face actually became a shade or two more red. Of the face of soil physicist-administrator Omer J. Kelley across the poker table, raising the bet in outrageous fashion and who, for a number of years, held the career fate of numerous western soil physicists in his hands as a USDA-ARS supervisor, and who had himself earned a part of his way through college as a dealer in a Reno gambling hall. Of a friendly but vigorous battle of words at a Western Regional Research conference between Sterling Taylor and Doral Kemper over some fine issue of thermodynamics and of the rubber daggers presented to each of them at subsequent banquet. Of the weighty reports carried to regional conferences by Sterling Taylor and carried home by other participants as overweight baggage on the airplane, but which demonstrated the prodigious amount of work, almost unparalleled by a professor at the time, carried on at the Utah station under Taylor. Of the finesse used by Kees van Bavel as he so carefully and meticulously takes apart some weak scientific assertions carelessly made by an enthusiastic but bumbling paper presenter. Of a luncheon at the home of Ernest Childs in Cambridge, rushed so as not to miss the rowing matches on the Thames. Of the unofficial and erratic membership of the "chow and marching club" formed by Ed Miller and/or Bob Miller and others in an ad hoc manner, after a day of soil physics papers at the national meetings, which marched, single, double, or triple file to some restaurant where several tables could be pushed together and technical, philosophic, or humorous discussions might take place. Of vicarious participation in the eventful life of Charles Slichter as one reads his biography (Ingraham, 1972) or his own book, *Science in a Tavern*, alluding to the formation of the Royal Society of London (Slichter, 1966), and later experiences the false sensation of having been there and known him. Of a father who, after reading something written by his son, would probably respond with a question or comment such as "Can you prove or defend it?" or "Good, now cut it in half." Or of innumerable other such experiences, some vicariously, where someone has bothered to write it down.

References

Adamson, A. W. 1960. *Physical Chemistry of Surfaces*. John Wiley, New York Chapter 1.

Akopov, P. I. 1935. The dynamics of soil moisture. *Pochvovedenie (Soil Science)* 4:584–592.

Allison, L. E. 1952. Effect of synthetic polyelectrolytes on the structure of saline and alkali soils. *Soil Sci.* 73:443–454.

Allison, L. E., and D. C. Moore. 1956. Effect of VAMA and HPAN soil conditioners in aggregation, surface crusting, and moisture retention in alkali soils. *Soil Sci. Soc. Am. Proc.* 20:143–146.

Allmaras, R. R., A. L. Black, and R. W. Rickman. 1973. Tillage, soil environment and root growth. In: *Proc. Natl. Cons. Tillage Conf.*, Soil Conservation Soc. Am., Ankeny, IA.

Alway, F. J., and V. L. Clark. 1916. Use of two indirect methods for the determination of the hygrosocpic coefficients of soils. *J. Agr. Res.* 7:345–359.

Alway, F. J., and G. R. McDole, and Guy R. McDole. 1917. Relation of movement of water in a soil to its hygroscopicity and initial moistness. *J. Agr. Res.* 10:391–428.

Anderson, A. B. C. 1943. A method of determining soil mositure content based on the variation of the electrical capacitance of soil, at low frequency, with moisture capacity. *Soil Sci.* 56:28–41.

Armbrust, D. V. and L. Lyles. 1975. Soil stabilizers to control wind erosion. In: W. C. Moldenhauer, W. R. Gardner, C. E. Clapp, M. M. Mortland, W. H. Gardner, and C. I. Rich, (eds.), *Soil Conditioners*, Spec. Publ. Ser., No. 7, Soil Sci. Soc. Am., Madison, WI, pp 77–82.

Ashton, F. M. 1956. Effects of a series of cycles of alternating low and high soil water contents on the rate of apparent photosynthesis in sugar cane. *Plant Physiol.* 31:266–274.

Aslyng, H. C. 1963. Soil physics terminology. *Int. Soc. Soil Sci. Bull.* 231:2–5.

Atanasiu, N. 1956. Eilhard Alfred Mitscherlich, 1974–1956. *Soil Sci.* 82:99–100.

Atterberg, O. T. 1912. Die Mechanische Bodenanalyse und die Klassification der Mineralboden Schwedens. *Int. Mitt. Bodenk.* 2:312–342.

Babcock, K. L., and R. Overstreet. 1955. Thermodynamics of soil moisture: a new application. *Soil Sci.* 80:257–263.

Barley, K. P., and E. L. Greacen. 1967. Mechanical resistance as a soil factor influencing the growth of roots and underground shoots. In: A. G. Norman (ed.) *Advance in Agronomy*, Vol. 19. Am. Soc. Agron., Madison, WI., pp. 1-43.

Baver, L. D. 1928. The relation of exchangeable cations to the physical properties of soils. *J. Am. Soc. Agron.* 20:921–941.

Baver, L. D. 1940. *Soil Physics.* John Wiley and Sons, New York (with subsequent editions in 1948, 1956, and, with Walter H. Gardner and Wilford R. Gardner, in 1972).

Baver, L. D., and G. M. Horner. 1933. Water content of soil colloids as related to their chemical composition. *Soil Sci.* 36:329–353.

Bear, F. E. 1962. On the ninetieth birthday of Sir E. John Russell. *Soil Sci.* 94:199–213.

Belcher, D. J., T. R. Cuykendall, and H. S. Sack. 1950. The Measurement of Soil Moisture and Density by Neutron and Gamma-Ray Scattering. Tech. Devel. Rept. 127, Technical Development and Evaluation Center, U.S. Civil Aeronautics Adm., Indianapolis, IN.

Bennett, H. H. 1939. *Soil Conservation*. McGraw-Hill Book Co., New York, 993 pp.

Bernhard, R. K., and M. Chasek. 1953. Soil density determination by means of radioactive isotopes. *Nondestruct. Test.* 11:17–23.

Betts, E. M. 1944. *Thomas Jefferson's Garden Book*. Am. Phil. Soc., Philadelphia, PA.

Betts, E. M. 1953. *Thomas Jefferson's Farm Book (with Commentary and relevant extracts from other writings)*. Am. Phil. Soc., Princeton Univ. Press, Princeton, NJ.

Biggar, J. W., and D. R. Nielsen. 1967. Miscible displacement and leaching phenomenon. In: R. M. Hagan, H. R. Haise, and T. W. Edminster (eds.), *Irrigation of Agricultural Lands*. Am. Soc. Agraon., Madison, WI, pp. 254–271.

Biggar, J. W., and D. R. Nielsen. 1976. Spatial variability of the leaching characteristics of a field soil. *Water Resour. Res.* 12:78–84.

Black, C. A., D. D. Evans, J. L. White, L. G. Ensminger, and F. E. Clark (eds.). 1965. *Methods of Soil Analysis*, Part I, Physical and Mineralogical Properties. Am. Soc. Agron., Madison, WI. 780 pp.

Blake, G. R. 1965. Bulk density. In: C. A. Black, D. D. Evans, J. L. White, L. G. Ensminger, and F. E. Clark (eds.) *Methods of Soil Analysis*. Am. Soc. Agron., Madison, WI, pp. 374–390.

Bloodworth, M. E., and J. B. Page. 1957. Use of thermistors for the measurement of soil moisture and temperature. *Soil Sci. Soc. Am. Proc.* 21:11–15.

Bodman, G. B., and E. A. Coleman. 1943. Moisture and energy conditions during downward entry of water into soils. *Soil Sci. Soc. Am. Proc.* 8:116–122.

Bodman, G. B., and N. E. Edlefsen. 1934. The soil moisture system. *Soil Sci.* 38:425–444.

Boersma, L. 1965. Field measurement of hydraulic conductivity above the water table. In: C. A. Black, D. D. Evans, J. L. White, L. G. Ensminger, and F. E. Clark (eds.). *Methods of Soil Analysis*. Am. Soc. Agron., Madison, WI, pp. 234–252.

Bolt, G. H., and Frissel, M. J. 1960. Thermodynamics of soil moisture. *Netherlands J. Agr. Sci.* 8:57–78.

Bolt, G. H., and R. D. Miller. 1958. Calculation of total and component potentials of water in soil. *Trans. Am. Geophys., Union* 39:917–928.

Bouma, J. 1973. Use of physical methods to expand soil survey interpretations of soil drainage conditions. *Soil Sci. Soc. Am. Proc.* 37:413–421.

Bouma, J., and J. L. Anderson. 1973. Relationships between soil structure characteristics and hydraulic conductivity. In: R. R. Bruce, K. W. Flach, and H.

M. Taylor (eds.), *Field Soil Water Regime*. SSSA Spec. Publ. Ser. No. 5, Soil Sci. Soc. Am., Madison, WI.

Bouyoucos, G. J. 1917. Measurement of the inactive, or unfree, moisture in the soil by means of the dilatometer method. *J. Agri. Res.* 8:195–217.

Bouyoucos, G. J. 1921. A new classification of soil moisture. *Soil Sci.* 11:33–48.

Bouyoucos, G. J. 1927a. The hydrometer as a new and rapid method for determining the colloidal content of soils. *Soil Sci.* 23:319–330.

Bouyoucos, G. J. 1927b. The hydrometer as a new method for the mechanical analysis of soils. *Soil Sci.* 23:343–349.

Bouyoucos, G. J. 1947. A new electrical resistance thermometer for soils. *Soil Sci.* 63:291–298.

Bower, C. A. 1972. In recognition of L. A. Richards on the occasion of his 68th birthday. *Soil Sci.* 113:229–231.

Bradfield, R. 1925. The chemical nature of colloidal clay. *J. Am. Soc. Agron.* 17:253–270.

Bradfield, R., and V. C. Jamison. 1939. Soil structure—attempts at its quantitative characterization. *Soil Sci. Soc. Am. Proc.* 3:70–76.

Breazeale, E. L., and W. T. McGeorge. 1951. Movement of water vapor in soils. *Soil Sci.* 71:181–185.

Breazeale, E. L., and W. T. McGeorge. 1953. Exudation pressure in roots of tomato plants under humid conditions. *Soil Sci.* 75:293–298.

Bresler, E. 1977. Trickle-drip irrigation: principles and application to soil-water management. In: N. C. Brady (ed.), *Advances in Agronomy*, Vol 29. Am. Soc. Agron., Madison, WI., pp. 343–393.

Bresler, E., W. D. Kemper, and R. J. Hanks. 1969. Infiltration, redistribution and subsequent evaporation of water from soil as affected by wetting rate and hysteresis. *Soil Sci. Soc. Am. Proc.* 33:832–840.

Briggs. L. J. 1897. The Mechanics of Soil Moisture. Bull. 10, USDA Bur. Soils, Washington, D.C.

Briggs, L. J. 1950. Limiting negative pressure of water. *J. Appl. Phys.* 21:721–722.

Briggs, L. J., and J. W. McLane. 1907. The Moisture Equivalent of Soils. USDA Bur. Soils, Bull. 45, Washington, D.C.

Briggs, L. J., and H. L. Shantz. 1912. The Wilting Coefficient for Different Plants and Its Indirect Determination. Bull. 230, USDA Bur. Plant Ind., Washington, D.C.

Brink, W. 1947. Hugh Hammond Bennett. *Soil Sci.* 64:256–257.

Browning, G. M. 1937. Changes in the erodibility of soils brought about by the application of organic matter. *Soil Sci. Soc. Am. Proc.* 2:85–96.

Bruce, R. R. 1972. Hydraulic conductivity evaluation of the soil profile from soil water retention relations. *Soil Sci. Soc. Am. Proc.* 36:555–561.

Bruce, R. R., and A. Klute. 1956. The measurement of soil moisture diffusivity. *Soil Sci. Soc. Am. Proc.* 20:458–462.

Bruce, R. R., and F. O. Whisler. 1973. Infiltration of water into layered field soils. In: A. D. Hadas, D. Swartzendruber, P. E. Rijtema, M. Fuchs, and B. Yaron (eds.), *Physical Aspects of Soil Water and Salts in Ecosystems*. Springer-Verlag, New York.

Bruce, R. R., K. W. Flach, and H. M. Taylor (eds.). 1973. *Field Soil Water Regime.* Spec. Publ. Series, Soil Sci. Soc. Am., Madison, WI.

Buckingham, E. 1904. Contributions to Our Knowledge of the Aeration of Soils. Bull. 25, USDA Bur. Soils, Washington, D.C.

Buckingham, E. 1907. Studies on the Movement of Soil Moisture. Bull. 38, USDA Bur. Soils, Washington, D.C.

Buckingham, W. 1921. On plastic flow through capillary tubes. *Proc. Am. Soc. Test. Mater.*, 21:1154–1161.

Buehrer, T. F. 1951. Frank Breazeale, 1874–1950. *Soil Sci.* 71:251–252.

Buol, S. W., F. D. Hole, and R. J. McCracken. 1973. *Soil Genesis and Classification.* Iowa State Press, Ames, IA.

Burr, W. W. 1913. The Storage and Use of Soil Moisture. Bull. 5, Nebraska Agr. Exp. Sta., Lincoln, NE.

Campbell, G. S. 1974. A simple method for determining unsaturated conductivity from moisture retention data. *Soil Sci.* 117:311–314.

Campbell, G. S., and W. H. Gardner. 1971. Psychrometric measurement of soil water potential: temperature and bulk density effects. *Soil Sci. Soc. Am. Proc.* 35:8–12.

Campbell, G. S., W. D. Zollinger, and S. A. Taylor. 1966. Sample changer for the thermocouple psychrometers: construction and some applications. *Agron. J.* 58:315–318.

Carr, C. E., and D. J. Greenland. 1975. Potential application of polyvinyl acetate and polyvinyl alcohol in the structural improvement of sodic soils. In: W. C. Moldenhauer, W. R. Gardner, C. E. Clapp, M. M. Mortland, W. H. Gardner, and C. I. Rich (eds.), *Soil Conditioners* Spec. Publ. Ser. 7, Soil Sci. Soc. Am., Madison, WI, pp. 1–12.

Cary, J. W. 1965. Water flux in moist soil: Thermal versus suction gradients. *Soil Sci.* 100:168–175.

Cary, J.W. 1967. The drying of soil: Thermal regimes and ambient pressures. *Arg. Meteorol.* 4:353–365.

Cary, J. W., and D. D. Evans (eds.). 1974. Soil Crusts. Tech. Bull. 214, Agr. Exp Sta., Univ. Arizona, Tucson, AZ.

Cary, J. W., and S. A. Taylor. 1967. The dynamics of soil water. Part II—Temperature and solute effects. In: R. M. Hagan, H. R. Haise, and T. W. Edminster (eds.), *Irrigation of Agricultural Lands.* Am. Soc. Agron., Madison, WI, pp. 245–251.

Chaptal, Count M. 1845. *Chymistry Applied to Agriculture* (Referred to in Jenny, 1961).

Chepil, W. S. 1943. Relation of wind erosion to water stable dry clod structure of soil. *Soil Sci.* 55:275–287.

Chepil, W. S. 1953. Factors that influence clod structure and erodibility of soil by wind. *Soil Sci.* 75:473–483.

Childs, E. C. 1936. Transport of water through heavy clay soils. I. *J. Agr. Sci.*, 26:392–405.

Childs, E. C. 1964. The ultimate moisture profile during infiltration in a uniform soil. *Soil Sci.* 97:173–178.

Childs, E. C. 1967. Soil moisture theory. *Hydrosci.* 4:73–117.

Childs, E. C. 1969. *The Physical Basis of Soil Water Phenomena.* John Wiley, London, 493 pp.

Childs, E. C., and N. Collis-George. 1948. Soil geometry and soil-water equilibrium. *Discuss. Faraday Soc.* 3:78–85.

Childs, E. C., and N. Collis-George. 1950a. The permeability of porous materials. *Proc. R. Soc., London* A201:392–405.

Childs, E. C., and N. Collis-George. 1950b. Movement of moisture in unsaturated soils. *Trans. Int. Congr. Soil Sci., Amsterdam* I-1–4

Cline, M. G. 1973. In recognition of Richard Bradfield on the occasion of his seventy-seventh birthday. *Soil Sci.* 115:273–275.

Cochrane, R. C. 1966. *Measures for progress, a history of the National Bureau of Standards.* Mimeograph, U.S. Dept. Commerce, Washington, D. C.

Coleman, E. A. 1944. The dependence of field capacity upon the depth of wetting of field soils. *Soil Sci.* 58:43–50.

Collis-George, N. 1955. Hysteresis in moisture content-suction relationships in soils. *Proc. Nat. Acad. Sci. India* 24A:80–85.

Collis-George, N. 1974. A laboratory study of infiltration-advance. *Soil Sci.* 117:282–287.

Conrad, J. P., and F. J. Veihmeyer. 1929. Root development and soil moisture. *Hilgardia* 4:113–134.

Cook, R. J., and R. I. Papendick. 1970. Effect of soil water on microbial growth, antagonism, and nutrient availability in relation to soil-borne fungal diseases of plants. In: T. A. Toussoun, R. V. Bega, and P. E. Nelson (eds.), *Root Diseases and Soil Borne Pathogens.* Univ. California Press, Berkeley, CA, pp. 81–88.

Corey, A. T. and W. D. Kemper. 1961. Concept of total potential in water and its limitations: a critique. *Soil Sci.* 91:199–302.

Corey, A. T., and A. Klute. 1985. Application of potential concept to soil water equilibrium and transport, Reviews of Research. *Soil Sci. Soc. Am. J.* 49:3–11.

Cowan, I. R. 1965. Transport of water in the soil-plant- atmosphere system. *J. Appl. Ecol.* 2:221–239.

Dalton, J. 1793. *Meteorological Observations and Essays.* W. Richardson, J. Phillips, and W. Pennington, London.

Danielson, R. E. 1971. Nutrient supply and uptake in relation to soil physical conditions. In: D. Hillel (ed.), *Optimizing the Soil Physical Environment Toward Greater Crop Yields.* Academic Press, New York, pp. 193–221.

Darcy, H. 1856. *Les Fontaines Publiques de la Ville de Dijon.* Dalmont, Paris.

Darwin, C. 1881. *The Formation of Vegetable Mold.* J. Murray, London.

Daumas, M. 1958. The chemistry of principles. In: R. Taton (ed.), *The Beginnings of Modern Science.* Thames and Hudson, London, pp. 322–323.

Davy, Sir Humphrey. 1813. *Elements of Agricultural Chemistry.* Longman, Hurst, Rees, Orme, and Brown, London.

Davidson, J. M., and D. D. Evans. 1960. Turbidimeter technique for measuring the stability of soil aggregates in water-glycerol mixture. *Soil Sci. Soc. Am. Proc.* 24:75–79.

Davidson, J. M., J. W. Biggar, and D. R. Nielsen. 1963. Gamma attenuation for measuring bulk density and transient water flow in porous materials. *J. Geophys. Res.* 68:4777–4783.

Day, P. R. 1942. The moisture potential of soils. *soil Sci.* 54:391–400.

Day, P. R. 1950. Physical basis of particle size analysis by the hydrometer method. *Soil Sci.* 70:363–374.

Day, P. R. 1953. Experimental confirmation of hydrometer theory. *Soil Sci.* 75:181–186.
Day, P. R. 1965. Particle fractionation and particle-size analysis. In: C. A. Black, D. D. Evans, J. L. White, L. G. Ensmiger, and F. E. Clark (ed's.), Methods of Soil Analysis. Part I. *Am. Soc. Agron.*, Madison, WI, pp. 454–567.
De Boodt, M. 1972. Improvement of soil structure by chemical means. In: D. Hillel (ed.), *Optimizing the Soil Physical Environment Towards Greater Crop Yield.* Academic Press, New York, pp. 43–55.
De Boodt, M. 1975. Use of soil conditioners around the world. In: W. C. Moldenhauer, W. R. Gardner, C. E. Clapp, M. M. Mortland, W. H. Gardner, and C. I. Rich (eds.), *Soil Conditioners* Spec. Publ. Ser. 7, Soil Sci. Soc. Am., Madison, WI, pp. 1–12.
De Bano, L. F. 1975. Infiltration, evaporation, and water movement as related to water repellency. In: W. C. Moldenhauer, W. R. Gardner, C. E. Clapp, M. M. Mortland, W. H. Gardner, and C. I. Rich (eds.), *Soil Conditioners.* Spec. Publ. Ser., No. 7, Soil Sci. Soc. Am., Madison, WI, pp. 155–164.
De Bano, L. F., and J. Letey (eds.). 1969. *Water-repellent soils.* Symp. Proc., May 6–10, Univ. Calif., Riverside, 354 pp.
De Leenheer, L. 1971. The influence of weather, crop and sampling depth on the measurement of pore size distribution in the arable layer of some cultivated silt soils. *Soil Sci.* 112:89–99.
De Leenheer, L., and M. De Boodt. 1959. Practical importance of the use of soil conditioners such as Krilium in controlling the degradation of soil structure. *Proc. Int. Symp. Soil Strut.* 1958:89–96.
Denmead, O. T., and R. H. Shaw. 1962. Availability of water to plants as affected by soil moisture content and meteorological conditions. *Agron. J.* 54:385–390.
Derjagin, B. V., M. K. Melnikova, and S. V. Nerpin. 1956. Theory of equilibrium and migration of soil moisture at various degrees of wetting. Rept. Comm I, 6th. Congress, Int. Soc. Soil Sci.
De'sigmond, Alexius A. J. 1935. Development of soil science. *Soil Sci.* 40:77–87.
De Vries, D. A. 1963. Thermal properties of soils. In: W. R. van Wijk (ed.). *Physics of Plant Environment.* North Holland, Amsterdam, pp. 210–235.
De Wiest, R. J. M. (ed.). 1969. *Flow Through Porous Media.* Academic Press, New York, 530 pp.
Dies, E. J. 1949. *Titans of the Soil: Great Builders of Agriculture.* Univ. North Carolina Press, Chapel Hill, NC.
Dobeneck, A. F. 1892. von Untersuchungen uber das Absorptionsvermogen und die Hygroskipizitat der Bodenkonstituenten. *Forsch. Geb. Agri.-Phyysik.* XI:163–228.
Duley, F. L. 1939. Surface factors affecting the rate of intake of water by soils. *Soil Sci. Soc. Am. Proc.* 4:60–64.
Edlefsen, N. E. 1933. A glass wool cell for measuring aqueous vapor pressure. *Rev. Sci. Instrum.* 4:345–346.
Edlefsen, N. E. 1934. A new method of measuring the aqueous vapor pressure of soils. *Soil Sci.* 38:29–35.
Edlefsen, N. E., and A. B. C. Anderson. 1943. Thermodynamics of soil moisture. *Hilgardia* 15:31–298.
Elrick, D. E. 1963. Unsaturated flow properties of soils. *Austr. J. Soil Res.* 1:1–8.
Emerson, W. W. 1965. Synthetic soil conditioners. *J. Agr. Sci.* 47:117–121.

Emerson, W. W., R. D. Bond, and A. R. Dexter (eds.). 1978. *Modification of Soil Structure.* John Wiley, Chichester, 438 pp.

Evans, D. D., and S. W. Buol. 1968. Micromorphological study of soil crusts. *Soil Sci. Soc. Am. Proc.* 32:19–22.

Evans, D. D., and D. Kirkham. 1949. Measurement of air premeability of soils in situ. Soil Sci. Soc. Am. Proc. 14:65–73.

Evelyn, J. 1673. *Terra, a Philosophical Discourse of Earth.* The Royal Society of London.

Ewing, S. 1922. The movement of saturated water vapor through quartz flour. *Soil Sci.* 13:57–61.

Fancher, G. 1956. Henry Darcy--Engineer and benefactor of mankind. *J. Petrol. Technol.*, 8:12–14.

Ferguson, H., and W. H. Gardner. 1962. Water content measurement in soil columns by gamma ray absorption. *Soil Sci. Soc. Am. Proc.* 26:11–14.

Fisher, H. A. 1926. On the capillary forces of an ideal soil. *J. Agr. Sci.* 16:492–503.

Fletcher, J. E. 1940. A dielectric method for determining soil moisture. *Soil Sci. Soc. Am. Proc.* 4:84–88.

Fletcher, J. E. 1949. Some properties of water solutions that influence infiltration. *Am. Geophys. Union* 30:548–554.

Gardner, H. R. 1969. Relation of water applicaton to evaporation and storage of soil water. *Soil Sci. Soc. Am. Proc.* 33:192–196.

Gardner, R. 1937. A method for measuring the capillary tension of soil moisture over a wide moisture range. *Soil Sci.* 43:277–283.

Gardner, R. 1945. Some effects of freezing and thawing on the aggregation and permeability of dispursed soils. *Soil Sci.* 60:437–443.

Gardner, W. 1955. Relation of temperature to moisture tension of soil. *Soil Sci.* 79:257–265.

Gardner, W. 1919. Capillary moisture-holding capacity. *Soil Sci.* 7:319–324.

Gardner, W. 1920. The capillary potential and its relation to soil-moisture constants. *Soil Sci.* 10:357–359.

Gardner, W., and J. A. Widstoe. 1921. The movement of soil moisture. *Soil Sci.* 11:215–232.

Gardner, W., O. W. Israelsen, N. E. Edlefsen, and H. Clyde. 1922. The capillary potential function and its relation to irrigation practice. *Phys. Rev.* 20:196.

Gardner, W. H. 1962. How water moves in the soil, Part I. The basic concept, Part II. In the field. *Crops and Soils* 15:7–11. (Revised 1968 and 1979: How water moves in the soil. *Crops and Soils* 21:7–12 and 32:13–18.)

Gardner, W. H. 1965. Water content. In: C. A. Black, D. D. Evans, J. L. White, L. E. Ensminger, and F. E. Clark (eds.), *Methods of Soil Analysis*, Part I. Am. Soc. Agron., Madison, WI, pp. 82–127.

Gardner, W. H. 1972. Use of synthetic soil conditioners in the 1950's and some implications to their further development. *Meded. Fac. Landbouwwet. Rijkuniv. Gent* 37:1046–1061.

Gardner, W. H. 1977. Historical highlights in American soil physics, 1776–1976. *Soil Sci. Soc. Am. J.* 41:221–229.

Gardner, W. H., and C. Calissendorff. 1967. Gamma-ray and neutron attenuation in measurement of soil bulk density and water content. Symp. on Use of Isotope and Radiation techniques in Soil Physics and Irrigation Studies, FAO,

IAEA, UNESCO, Istanbul, Turkey. Int. Atomic Energy Agency, Vienna, pp. -113.

Gardner, W. H., and W. Gardner. 1951. Flow of soil moisture in the unsaturated state. *Soil Sci. Soc. Am. Proc.* 15:42–50.

Gardner, W. H., G. S. Campbell, and C. Calissendorff. 1972. Systematic and random errors in dual gamma energy bulk density and water content measurements. *Soil Sci. Soc. Am. Proc.* 36:393–398.

Gardner, W. H., S. L. Rawlins, A. A. Rode, C. E. Kellogg, and T. J. Marshall. 1973. Hydro-physics of arid and irrigated soils. In: V. A. Kovda, C. van den Berg, and R. M. Hagan (eds.). *Irrigation, Drainage and Salinity.* Hutchinson/FAO/UNESCO, London.

Gardner, W. R. 1956. Calculation of capillary conductivity from pressure plate outflow data. *Soil Sci. Soc. Am. Proc.* 20:317–320.

Gardner, W. R. 1958. Some steady-state solutions of the unsaturated moisture flow equation with application to evaporation from a water table. *Soil Sci.* 85:228–232.

Gardner, W. R. 1959. Solutions of the flow equation for the drying of soils and other porous media. *Soil Sci. Soc. Am. Proc.* 23:183–187.

Gardner, W. R. 1960. Dynamic aspects of water availability to plants. *Soil Sci.* 89:63–73.

Gardner, W. R. 1967. Development of modern infiltration theory and application in hydrology. *Trans. Am. Soc. Agr. Eng.* 10:379–382, 390.

Gardner, W. R. 1972. The impact of L. A. Richards upon the field of soil water physics. *Soil Sci.* 113:232–237.

Gardner, W. R. 1974. The permeability problem. *Soil Sci.* 117:243–249.

Gardner, W. R., and D. I. Hillel. 1962. The relation of external evaporative conditions to the drying of soils. *J. Geophys. Res.* 67:4319–4325.

Gardner, W. R., and D. Kirkham. 1952. Determination of soil moisture by neutron scattering. *Soil Sci.* 73:391–401.

Gardner, W. R., D. Hillel, and Y. Benyamini. 1970. Post-irrigation movement of soil water. 1. Redistribution. *Water Resour. res.* 6:851–861.

Glinka, K. D. 1931. Treatise on Soil Science. (Pochvovedenie), Transl. Israel Program for Sci. Transl., Jerusalem, NSF, Washington, D. C.

Grable, A. R. 1966. *Soil Aeration and Plant Growth, Adv. Agron*, In: A. G. Norman (ed.) Vol. 18. New York, pp. 47–106.

Green, R. E., and J. C. Corey. 1971. Calculation of hydraulic conductivity: a further evaluation of some predictive methods. *Soil Sci. Soc. Am. Proc.* 35:3–8.

Green, W. H., and G. A. Ampt. 1911. Studies on soil physics, Part I. Flow of air and water through soils. *J. Agr. Sci.* 4:1–24.

Green, W. H., and G. A. Ampt. 1912. Studies on soil physics, Part II. Permeability of an ideal soil to air and water. *J. Agr. Sci.* 5:1–26.

Greenland, D. J. 1965. Interaction between clays and organic compounds in soils: Part I and Part II. *Soils Fertil.* 28:415–425; 28:521–532.

Groenevelt, P. H., and G. H. Bolt. 1969. Non-equilibrium thermodynamics of the soil-water system. *J. Hydrol.* 7:358–388.

Groenevelt, P. H., and J. Y. Parlange. 1974. Thermodynamic stability of swelling soils. *Soil Sci.* 118:1–5.

Gurr, C. G. 1947. Freezing point of soil water in relation to permanent wilting percentage. *Austr. Council Sci. Indust. Res. J.* 20:105–109.

Gurr, C. G. 1962. Use of gamma rays in measuring water content and permeability in unsaturated soil columns. *Soil Sci.* 94:224–229.

Gurr, C. G., T. J. Marshall, and J. T. Hutton. 1952. Movement of water in soil due to a temperature gradient. *Soil Sci.* 74:335–345.

Hadas, A., D. Swartzendruber, P. E. Rijtema, M. Fuchs, and B. Yaron, (eds.). 1973. *Physical Aspects of Soil Water and Salts in Ecosystems*, Proc. 1971 Symposium, "Soil water physics and technology," Rehovot, Israel. Springer-Verlag, New York.

Hagan, R. M., H. R. Haise, and T. W. Edminster (eds.). 1967. *Irrigation of Agricultural Lands.* Am. Soc. Agron., Madison, WI.

Hagen, G. H. L. 1839. Ueber die Bewegung des Wassers in engen cylindrischen Rohren. *Poggend. Annal.* 46:423–442.

Haines, W. B. 1925a. Studies in the physical properties of soils. I. Mechanical properties concerned in cultivation. *J. Agr. Sci.* 15:178–200.

Haines, W. B. 1925b. Studies in the physical properties of soils. II. A note on the cohesion developed by capillary forces in an ideal soil. *J. Agr. Sci.* 15:529–535.

Haines, W. B. 1925c. Studies in the physical properties of soils. III. Observations on the electrical conductivity of soils. *J. Agr. Sci.* 15:536–543.

Haines, W. B. 1927. Studies in the physical properties of soils. IV. A further contribution to the theory of capillary phenomena in soil. *J. Agr. Sci.* 17:264–290.

Haines, W. B. 1930a. Studies in the physical properties of soils. V. The hysteresis effect in capillary properties and the modes of moisture distribution associated therewith. *J. Agr. Sci.* 20:97–116.

Haines, W. B. 1930b. On the existence of two equilibrium series in soil capillary phenomena. Proc. 2nd. Int. Congr. Soil Sci., Leningrad, 1930, Comm. 1 Soil Physics, 8–13.

Haise, H. R., and R. M. Hagen. 1967. Soil, plant, and evaporative measurements as criteria for scheduling irrigation. In: R. M. Hagan, H. R. Haise, and T. W. Edminster (eds.), *Irrigation of Agricultural Lands.* Am. Soc. Agron., Madison, WI, pp. 577–597.

Haise, H. R., and O. J. Kelley. 1950. Causes of dirunal fluctuations of tensiometers. *Soil Sci.* 70:301–313.

Hall, W. A. 1956. An analytical derivation of the Darcy equation. *Trans. Am. Geophys. Union*, 37:185–188.

Halley, E. 1687. An estimate of the quantity of vapour raised out of the sea by the warmth of the sun; derived from an experiment shown before the Royal Society of London at one of their later meetings. *Phil. Trans. R. Soc. London* 16:366–370.

Halley, E. 1691. On the circulation of the vapours of the sea and the origin of springs. *Phil. Trans. R. Soc. London* 17:468–473.

Hanks, R. J., H. S. Jacobs, H. E. Schimmelpfennig, and M. Nimah. 1973. Evaluation of several methods for estimating evapotranspiration under semiarid conditions. In: A. Hadas, D. Swartzendruber, P. E. Rijtema, M. Fuchs, and B. Yaron (eds.), *Physical Aspects of Soil Water and Salts in Ecosystems.* Springer-Verlag, New York.

Harris, J. A. 1915. On a criterion of substratum homogeneity (or heterogeneity) in field experiments. *Am. Nat.* 49:430–454.

Harris, J. A. 1920. Practical universality of field heterogeneity as a factor influencing plot yields. *J. Agr. Res.* 29:279–314.

Hedrick, R. M., and D. T. Mowry. 1952. Effect of synthetic polyelectroltyes on aggregation, aeration,and water relationships of soil. *Soil Sci.* 73:427–441.

Heinrich, R. 1894. Zweiter Bericht uber die Verhaltnisse and Wirksamkeit der landwirtschaflichen Versuchs-Stationen zu Rostock, p. 29 (cited by Briggs and Schantz, 1912).

Hele-Shaw, H. S. 1898. Stream-line Motion of a Viscous Film. I. Experimental Investigation of the Motion of a Thin Film of Viscous Fluid. Report of the 68th Meeting of British Assoc. Adv. Sci., Bristol. John Murray, London, pp. 136–142.

Hilgard, E. W. 1860. *Report in the Geology and Agriculture of the State of Mississippi.* State Printer, Jackson, MS. 391 pp., map.

Hilgard, E. W. 1892. A report on the relations of soil to climate. *USDA Weather Bur. Bull.* 3:1–59.

Hilgard, E. W. 1903. The Chemistry of soil as related to crop production. *Science* 18:755–760.

Hilgard, E. W. 1904. Soil management. *Science* 20:605–608.

Hilgard, W. W. 1906 (1921 ed.). *Soils, Their Formation, Properties, Composition, and Relations to Climate and Plant Growth in the Humid and Arid Regions.* The Macmillan Co., New York.

Hillel, D. 1971. *Soil and Water, Physical Principles and Processes.* Academic Press, New York, 288 pp.

Hillel, D. (ed.). 1972. *Optimizing the Soil Physical Environment Towards Greater Crop Yields.* Academic Press, New York.

Hillel, D., and W. R. Gardner. 1969. Steady infiltration into crust-topped profiles. *Soil Sci.* 108:137–142.

Hillel, D., and W. R. Gardner. 1970a. Transient infiltration into crust-topped profiles. *Soil Sci.* 109:69–70.

Hillel, D., and W. R. Gardner. 1970b. Measurement of unsaturated conductivity and diffusivity by infiltration through an impeding layer. *Soil Sci.* 109:149–153.

Holmes, J. W., and K. G. Turner. 1958. The measurement of water content of soils by neutron scattering: A portable apparatus for field use. *J. Agr. Eng. Res.* 3:199–204.

Holmes, J. W., S. A. Taylor, and S. J. Richards. 1967. Measurement of soil water. In: R. M. Hagan, H. R. Haise, and T. W. Edminster (eds.), *Irrigation of Agricultural Land.* Am. Soc. Agron., Madison, WI, pp. 275–298.

Hopkins, C. G. 1910. *Soil Fertility and Permanent Agriculture.* Ginn and Co., New York.

Horton, R. E. 1940. An approach toward a physical interpretation of infiltration capacity. *Soil Sci. Soc. Am. Proc.* 5:399–417.

Ingraham, M. H. 1972. *Charles Sumner Slichter--the Golden Vector.* Univ. Wisconsin Press, Madison, WI, 316 pp.

Ingvalson, R. D., J. D. Oster, S. L. Rawlins, and G. J. Hoffman. 1970. Measurement of water potential and osmotic potential in soil with a combined thermocouple and salinity sensor. *Soil Sci. Soc. Am. Proc.* 34:570–574.

International Society of Soil Science. 1974. Soil Physics Terminology. Bull. 44, Int. Soc. Sci., pp. 10–17.

Irmay, S. 1956. Extension of Darcy Law to Unsteady Unsaturated Flow Through Porous Media. 1956 Symposia Darcy, Publ. 41. Assoc. Intl. d'Hydrologic Scientifique, Coop. UNESCO Outerick, Louvain, Belgium, pp. 57–66.

Israelsen, O. W. 1927. The application of hydrodynamics to irrigation and drainage problems. *Hilgardia* 2:479–528.

Iwata, S. 1972. Thermodynamics of soil water: I. The energy concept of soil water. *Soil Sci.* 113:162–166.

Jackson, R. D. 1964. Water vapor distribution in relataively dry soil. I. Theoretical considerations and sorption experiments. *Soil Sci. Soc. Am. Proc.* 28:172–176.

Jackson, R. D., and F. D. Whisler. 1970. Equations for approximating vertical non-steady-state drainage of soil columns. *Soil Sci. Soc. Am. Proc.* 34:715–718.

Jamison, V. C. 1953. Changes in air-water relationships due to structural improvement of soils. *Soil Sci.* 76:143–151.

Jenny, H. 1961. *E. W. Hilgard and the Birth of Modern Soil Science.* Collana Della Riviste "Agrochimica," Pisa.

Johnson, S. W. 1856. Physical Properties of Soils as Affecting Soil Fertility. (Mentioned in *"Liberty Hyde Bailey,"* by Andrew Denny Rogers, III, 1949, Princeton Univ. Press.)

Johnson, S. W. 1877. *On the Reasons for Tillage.* Report, Conn. B. Agr., New Haven, CT., pp. 133–151.

Johnson, S. 1878. Experiments on the Relation of Soils to Water. First Report Conn Agr. Exp. Sta., New Haven, CT, pp. 83–102.

Jury, W. A., W. R. Gardner, P. G. Saffigna and C. B. Tanner. 1976. Model for predicting simultaneous movement of nitrate and water through a loamy sand. *Soil Sci.* 122:36–43.

Kastanek, K. F. 1971. Numerical simulation technique for vertical drainage from a soil column. *J. Hydrol.* 14:213–232.

Keen, B. A. 1914. The evaporation of water from soil. *J. Agr. Sci.* 6:456–475.

Keen, B. A. 1919. A quantitative relation between soil and the soil solution brought out by freezing-point determinations. *J. Agr. Sci.* 9:400–415.

Keen, B. A. 1924. On the moisture relationships in an ideal soil. *J. Agr. Sci.* 14:170–177.

Keen B. A. 1926. The physicist in agriculture with special reference to soil problems (with introduction by Sir Daniel Hall). *Physics in Industry*, 4:27.

Keen, B. A. 1928a. First commission soil mechanics and physics. *Soil Sci.* 25:9–21

Keen, B. A. 1928b. Mechanical analysis: national and international. *Soil Res.* 1:43–49.

Keen, B. A. 1931. *The Physical Properties of the Soil.* Longmans, Green and Co., London.

Keen, B. A., and J. R. H. Coutts. 1928. "Single value" soil properties: a study of the significance of certain soil constants. *J. Agr. Sci.* 18:740–765.

Keen, B. A., and H. Raczkowski. 1921. The relation between the clay content and certain physical properties of a soil. *J. Arg. Sci.* 11:441–449.

Keen, B. A., and E. J. Russell. 1921. The factors determining soil temperature. *J. Arg. Sci.* 11:211–239.

Keilhack, K. 1912. *Lehrbuch der Grundwasser--und Quellenkunde.* Verlag von Gebruder Borntraeger, Berlin.

Kelley, W. P. 1939. Charles Fredrick Shaw, 1881–1939. *Soil Sci.* 48:524–526.

Kellogg, C. E. 1938. Soil and society. In: H. G. Knight (ed.), *USDA Yearbook of Agriculture*. U.S. Gov. Printing Off. Washington, D.C.

Kellogg, C. E. 1963. Shifting cultivation. *Soil Sci.* 95:221-230.

Kemper, W. D. 1959. Estimation of osmotic stress in soil water from the electrical resistance of finely porous ceramic units. *Soil Sci.* 87:345-349.

Kemper, W. D. 1965. Aggregate stability. In: C. A. Black, D. D. Evans, J. L. White, L. G. Ensminger, and F. E. Clark (eds.), *Methods of Soil Analysis*, Part I. Am. Soc. Agron., Madison, WI, pp. 511-519.

Kemper, W. D., and E. J. Koch. 1966. Aggregate Stability of Soils from Western United States and Canada. Tech. Bull. 1355, USDA, Washington, D.C.

Kijne, J. W., and S. A. Taylor. 1964. Temperature dependence of soil water vapor pressure. *Soil Sci. Soc. Am. Proc.* 28:595-599.

King, F. H. 1982. Observations and Experiments on the Fluctuations in the Level and Rate of Movement of Ground Water on the Wisconsin Agricultural Experiment Station Farm at Whitewater, Wisc. Bull. 5, U.S. Weather Bur., Washington D.C.

King, F. H. 1895. *The Soil*. The Macmillan Co., New York.

King, F. H. 1898. Principles and Conditions of the Movements of Ground Water. U.S. Geol. Surv. 19th Ann. Rept. Pt. 2, pp. 59-294.

King, F. H. 1899. *Irrigation and Drainage*. The Macmillan Co., New York.

King, F. H. 1904. "Investigations in Soil Management. being three of six papers on the influence of soil management upon the water-soluble salts in soils, and the yeild of crops. Published by the author, Madison, WI.

King, F. H. 1905. Investigations in Soil Management. Bull. 26, USDA, Bur. Soils, Washington, D.C.

Kirkham, D. 1940. Pressure and streamline distribution in waterlogged land overlying an impervious layer. *Soil Sci. Soc. Am. Proc.* 5:65-68.

Kirkham, D. 1961. Soil Physics. 1936-61 and a look ahead. *Soil Sci. Soc. Am. Proc.* 25:423-427.

Kirkham, D. 1972. Problems and trends in drainage research, mixed boundary conditions. *Soil Sci.* 113:285-293.

Kirkham, D., and C. L. Feng. 1949. Some tests of the diffusion theory, and laws of capillary flow, in soils. *Soil Sci.* 67:29-39.

Kirkham, D., and W. C. Powers. 1972. *Advanced Soil Physics*. John Wiley, New York.

Klute, A. 1952a. A numerical method for solving the flow equation for water in unsaturated materials. *Soil Sci. 73:105-116.*

Klute, A. 1952b. Some theoretical aspects of the flow of water in unsaturated soils. Soil Sci. Soc. Am. Porc. 16:144-148.

Klute, A. 1972. The determination of the hydraulic conductivity and diffusivity of unsaturated soils. *Soil Sci.* 113:264-276.

Klute, A., and L. A. Richards. 1962. Effect of temperature on relative vapor pressure of water in soil. Apparatus and preliminary measurements. *Soil Sci.* 93:391-397.

Kohnke, H. 1946. The practical use of the energy concept of soil moisture. *Soil Sci. Soc. Am. Proc.* 11:64-66.

Kohnke, H. 1968. *Soil Physics*. McGraw-Hill, New York, 224 pp.

Korven, H. C., and S. A. Taylor. 1959. The Peltier effect and its use for determining relative activity of soil water. *Can. J. Soil Sci.* 39:76-85.

Kostiakov, A. N. 1932. On the dynamics of the coefficient of water percolation in soils and their necessity of studying it from the dynamic point of view for the purposes of amelioration. In: *Trans. 6th Cong. Int. Soc. Soil Sci.* pp. 17–21.

Kovda, V. A., C. Van Den Berg, and R. M. Hagan. 1973. *Irrigation, Drainage and Salinity, an International Source Book.* Hutchinson/FAO/UNESCO, London.

Kozlowski, T T. (ed.). 1968, 1976. *Water Deficits and Plant Growth*, Vols. I and IV. Academic Press, New York.

Kramer, P. J. 1932. The absorption of water by root systems of plants. *Am. J. Bot.* 19:148–164.

Kramer, P. J. 1949. *Plant and Soil Water Relationships.* McGraw-Hill, New York, 347 pp.

Kramer, S. N. 1958. *History Begins at Sumer.* Doubleday, Garden City, New York, 247 pp.

Kramer, S. N. 1963. *The Sumerians, their History, Culture, and Character.* Univ. Chicago Press, Chicago, 355 pp.

Kuipers, H. 1970. Introduction: Historical notes on the zero-tillage concept. *Netherlands J. Agr. Sci.* 18:219–224.

Kunze, R. J., and D. Kirkham. 1962. Simplified accounting for membrane impedance in capillary conductivity measurements. *Soil Sci. Soc. Am. Proc.* 26:421–430.,

Kunze, R. J., and D. Kirkham. 1964. Capillary difusion and self-diffusion of soil water. *Soil Sci.* 97:145–151.

Lamb, H. 1959 (6th ed.). *Hydrodynamics.* Dover Publications, New York.

Langmuir. I. 1918. The absorption of gases on plane surfaces of glass, mica and platinum. *J. Am. Chem. Soc.* 40:1361–1403.

LaRocque, A. 1957. *The Admirable Discourses of Bernard Palissy* (with biographical introduction). Univ. Illinois Press, Urbana, 264 pp.

Lebedeff (Lebedev), A. F. 1927. The movement of ground and soil waters. *Proc. 1st Int. Cong. Soil Sci.* 1:459–494.

Lebedev A. F. 1918. *Die Bewegung des Wassers im Boden und im Untergrund.* Berlin. (English transl's., Soil And Ground Waters, by W. C. Lowdermilk in 1929, J. F. Lutz in 1933. Listed in Nat. Union Cat. Pre-1956 Imprints. Lib. Cong. Printed Cards, Vol. 32, Mansell, 1974.)

Lemon, E. R. 1962. Soil aeration and plant relations. I. Theory. *Agron. J.* 54:176–170.

Lemon, E. R. 1968. Energy and water balance of plant communities. In: T. T. Kozlowski (ed.), 1968, *Water Deficits and Plant Growth*, Vol. 1, Academic Press, New York.

Lemon, E., and C. C. Wiegand. 1962. Soil aeration and plant root relations. II. Root respiration. *Agraon. J.* 54:171–175.

Leroux, D. 1931. A.-Th. Schloesing (1856–1930) sa vie et son oeuvre. *Agronomy.* 1:3–69.

Letey, J. 1975. The use of nonionic surfactants on soils. In: W. C. Moldenhauer, W. R. Gardner, C. E. Clapp, M. M. Mortland, W. H. Gardner, and C. I. Rich (eds.) *Soil Conditioners.* Spec. Publ. Ser. No. 7, Soil Sci. Soc. Am., Madison, WI, pp. 145–154.

Lewis, M. R. 1937a. Rate of Flow of Capillary Moisture. Tech. Bull. 579, USDA, Washington, D.C.

Lewis, M. R. 1937b. The rate of infiltration water in irrigation practice. *Eos Trans. AGU*, 18:361–368.
Linford, L. B. 1926. The relation of light to soil moisture phenomena. *Soil Sci.* 22:233–252.
Lipman, J. G. 1916. Introductory. *Soil Sci.* 1:3–4.
Lipman, J. G. 1928. History of the organization of the International Society of Soil Science. *Soil Sci.* 25:3–21.
Livingston, B. E., and L. A. Hawkins. 1915. The Water-Relation between Plant and Soil. Publ. 204, Carnigie Inst., Washington, D. C., pp. 5–48.
Livingston, B. E., J. C. Britton, and F. R. Reid. 1905. Studies on the Properties of Unproductive Soils. Bull. 28, U.S. Bur. Soils, Washington, D.C.
Loughridge, R. H. 1892–1894. Investigations in soil physics: The Capillary Rise of Water in Soils. Calif. Exp. Sta. Ann. Rept., Berkeley, CA, pp. 91–100.
Low, A. J. 1954. A study of soil structure in the field and in the laboratory. *J. Soil Sci.* 5:57–74.
Low, P. F. 1961a. Concept of total potential in water and its limitations: a critique. *Soil Sci.* 91:303–305.
Low, P. F. 1961b. Physical chemistry of clay-water interaction. *Adv. Agron.* 13:269–327.
Low, P. F., and D. M. Anderson. 1958. Osmotic pressure equations for determining thermodynamic properties of soil water. *Soil Sci.* 86:251–253.
Lundegardh, H. 1934. Sven Oden, 1888–1934. *Soil Sci.* 37:429.
Luthin, J. N., and D. Kirkham. 1949. A piezometer method for measuring permeability of soil *in situ* below a water table. *Soil Sci.* 68:349–358.
Lutz, J. F. 1952. Mechanical impedance and plant growth. In: B. T. Shaw (ed.), *Soil Physical Conditions and Plant Growth.* Academic Press, New York, pp. 43–67.
Lutz, J. F. 1977. History of the Soil Science Society of America. *Soil Sci. Soc. Am. J.* 41:152–173.
Lynde, C. J. 1912. Osmosis in soils: soils act as semipermeable membranes. *J. Am. Soc. Agron.* 4:102–108.
Lynde, C. J., and F. W. Bates. 1912. Further studies in the osmosis of soils. *J. Am. Soc. Agron.* 4:109–121.
Lynde, C. J., and H. A. Dupre. 1913. On a new method of measuring the capillary lift of soils. *J. Am. Soc. Agron.* 5:107–116.
Lyon, T. L. 1933. History of the organization of the American Society of Agronomy. *J. Am. Soc. Agron.* 25:1–9.
MacCurdy, E. (ed., transl.). 1938. *The Notebooks of Leonardo da Vinci.* Reynal and Hitchcock, New York.
Marei, S. M. 1974. A Hele-Shaw model study of oscillating water tables in drained homogeneous soils. *Soil Sci.* 117:301–305.
Marshall, T. J. 1958. A relation between permeability and size distribution of pores. *J. Soil Sci.* 9:1–8.
Marshall, T. J. 1959. Relations between Water and Soil. Tech. Comm. 50, Commonwealth Bureau of Soils, Harpenden, Farnham Royal, Bucks., England, 91 pp.
Marshall, T. J., and C. J. Gurr. 1954. Movement of water and chlorides in relatively dry soil. *Soil Sci.* 77:147–152.

Marshall, T. J., and J. W. Holmes. 1979. *Soil Physics.* Cambridge Univ. Press, Cambridge.

Martin, W. P., G. S. Taylor, J. C. Engibous, and E. Burnett. 1952. Soil and crop responses from field applications of soil conditioners. *Soil Sci.* 73:455–471.

Mason, D. D. 1948. Guy Wollard Conrey, 1887–1958. *Soil Sci.* 66:171–172.

Mayer, E. A. 1905, *Agriculturchemie* (cited by Hilgard, 1906).

McCall, A. G. 1909. Instruction in soil physics. *Proc. Am. Soc. Agron.* 1:207–211.

McCalla, T. M. 1944. Water-drop method of determining stability of soil structure. *Soil Sci.* 58:117–121.

McHenry, J. R., and M. B. Russell. 1943. Elementary mechanics of aggregation of puddled materials. *Soil Sci. Soc. Am. Proc.* 8:71–78.

Mehuys, G. R., L. H. Stolzy, and J. Letey. 1975. Temperature distribution under stones submitted to a diurnal heat wave. *Soil Sci.* 120:437–441.

Meinzer, O. E. 1934. The history and development of ground-water hydrology. *Washington Acad. Sci. J.* 24:6–32.

Middleton, H. E. 1920. The moisture equivalent in relation to the mechanical analysis of soils. *Soil Sci.* 9:159–167.

Miller, D. E., and W. H. Gardner. 1962. Water infiltration into stratified soil. *Soil Sci. Soc. Am. Proc.* 26:115–119.

Miller, E. E., and D. E. Elrick. 1958. Dynamic determination of capillary conductivity extended from non-negligible membrane impedance. *Soil Sci. Soc. Am. Proc.* 22:483–486.

Miller, E. E., and A. Klute. 1967. The dynamics of soil water. Part I. Mechanical Forces. In: R. M. Hagan, H. R. Haise, and T. W. Edminster (eds.), *Irrigation of Agricultural Lands.* Am. Soc. Agron., Madison, WI, pp. 209–240.

Miller, E. E., and R. D. Miller. 1956. Physical theory for capillary flow phenomena. *J. Appl. Phys.* 27:324–332.

Miller, R. D. 1951. A technique for measuring soil-moisture tensions in rapidly changing systems. *Soil Sci.* 72:291–301.

Millington, R. J. and J. P. Quirk. 1959. Permeability of porous media. *Nature* (London), 183:387–388.

Millington, R. J., and J. P. Quirk. 1960. Transport in porous media. *Trans. 7th Int. Congr. Soil Sci.* I:97–106.

Mitscherlich, E. A. 1901. Untersuchungen uber die physikalischen Bodeneigenshaften. *Landw. Jahrb.*, 30:360–445.

Mitscherlich, E. A. 1905. *Bodenkunds für Land-und-Forstwirthe.* P. Parey, Berlin.

Mitscherlich, E. A. 1930. *Die Bestimmung des Dungebedürfnisses des Bodens.* P. Parey, Berlin.

Moldenhauer, W. C., and W. D. Kemper. 1964. Interdependence of water drop energy and clod size in infiltration and clod stability. *Soil Sci. Soc. Am. Proc.* 33:297–301.

Moldenhauer, W. C., W. R. Gardner, C. E. Clapp, M. M. Mortland, W. H. Gardner, and C. I. Rich (eds.). 1975. *Soil Conditioners.* Spec. Publ. Ser. No. 7, Soil Sci. Soc. Am., Madison, WI.

Moltz, F. J. 1981. Models of water transport in the soil-plant system: a review. *Water Resour. Res.* 17:1245–1260.

Monteith, J. L., and P. C. Owen. 1958. A thermocouple method for measuring relative humidity in the range 95–100%. *J. Sci. Instrum.* 35:443–446.

Moore, R. E. 1939. Water conduction from shallow water tables. *Hilgardia*, 12:383-426.

Moore, R. E. 1941. The relation of soil temperature to soil moisture: pressure potential, retention, and infiltration rate. *Soil Sci. Soc. Am. Proc.* 5:61-64.

Morel-Seytoux, H. J. 1969. Introduction to flow of immiscible liquids in porous media. In: R. J. M. De Wiest (ed.), *Flow Through Porous Media*. Academic Press, New York, pp. 455-516.

Morel-Seytoux, H. J., and J. Khanji. 1974. Derivation of an equation of infiltration. *Water Resour. Res.*, 10:795-800.

Mosier, J. G., and A. F. Gustafson. 1917. *Soil Physics and Management*. J. B. Lippincott, Philadelphia.

Muir, A. 1950. Gilbert Wooding Robinson, 1889-1950. *Soil Sci.* 70:171-172.

Neal, O. R., L. A. Richards, and M. B. Russell. 1937. Observations on moisture conditions in lysimeters. *Soil Sci. Soc. Am. Proc.* 2:35-44.

Nerpin, S. V., and A. F. Chudnovskii. 1967 (Transl. 1970). *Physics of the Soil*. Transl. from Russian by IPST Staff. Israel Program for Scientific Translations, Jerusalem.

Nielsen, D. R., D. Kirkham, and W. R. Van Wijk. 1959. Measuring water stored temporarily above field moisture capacity. *Soil Sci. Soc. Am. Proc.* 23:408-412.

Nielsen, D. R., D. Kirkham, and E. R. Perrier. 1960. Soil capillary conductivity: Comparison of measured and calculated values. *Soil Sci. Soc. Am. Proc.* 24:157-160.

Nielsen, D. R., R. D. Jackson, J. W. Cary, and D. D. Evans, (eds.). 1972. *Soil Water*. Am. Soc. Agron., Madison, WI, 175 pp.

Nielsen, D. R., J. W. Biggar, and K. T. Erh. 1973. Spatial variability of field measured soil water properties. *Hilgardia*, 42:215-260.

Nofziger, D. L., and D. Swartzendruber. 1974. Material content of binary physical mixtures as measured with a dual-energy beam of gamma rays. *J. Appl. Physics*, 45:5443-5449.

Nye, P. H., and D. J. Greenland. 1960. The Soil Under Shifting Cultivation. Tech. Comm. 5, Commonwealth Bur. Soils, Harpenden, Commonwealth Agr. Bur., Farnham Royal, Bucks., England.

Oden, S. 1915. Eine neue Methode zur mechanischen Bodenanalyse. *Int. Mitt. Bodenk.* 5:257-311.

Oden, S. 1925. The size distribution of particles on soils and the experimental methods of obtaining them. *Soil Sci.* 19:1-35.

Osborne, T. B. 1887. The Methods of Mechanical Soil Analysis. Ann. Rep., Connecticut Agr. Exp. Sta., New Haven, pp. 143-162.

Page, J. B. 1948. Advantages of the pressure picnometer for measuring the pore space in soils. *Soil Sci. Soc. Am. Proc.* 12:81-84.

Palissy, B. 1563. *Recepte Veritable* (a short title for a long one given in LaRocque, 1957; containing a discussion "On various salts in agriculture" which appears again in "Admirable Discourses" below).

Palissy, B. 1580. *Admirable Discourses*. Translated by Aurele Larocque, 1957, as *The Admirable Discourses of Bernard Palissy* (with biographical introduction). Univ. Illinois Press, Urbana.

Papendick, R. I., and G. S. Campbell. 1974. Water potential in the rhizosphere and plant and methods of measurement and experimental control. In: G. W. Bruehl (ed.), *Biology and Control of Soil-Borne Plant Pathogens*. Am. Phytopathol. Soc., St. Paul, MN, pp. 39-49.

Parlange, J. Y. 1971. Theory of water-movement in soils: I. One-dimensional, and II. Two-dimensional infiltration. *Soil Sci.* 111:134–137, 170–174.

Patten, H. E. 1908. Heat Transference in Soils. Bull. 59, USDA Bur. Soils, Washington, D.C.

Patten, H. E., and F. E. Gallagher. 1908. Absorption of Vapors and Gases by Soils. Bull. 51, USDA Bur. Soils, Washington, D.C.

Peck, A. J. 1965a. Moisture profile development and air compression during water uptake by bounded porous bodies: 2. Horizontal columns. *Soil Sci.* 99:327–334.

Peck, A. J. 1965b. Moisture profile development and air compression during water uptake by bounded porous bodies: 3. Vertical columns. *Soil Sci.* 100:44–51.

Peck, A. J. 1966. Diffusivity determination by a new outflow method. In: P. E. Rijtema and H. Wassink (eds.), *Water in the Unsaturated Zone*, Vol. I. Publ. No. 82, Int. Assoc. Sci. Hydrol., UNESCO, Paris, pp. 191–202.

Peck, A. J. 1968. Theory of the Spanner psychrometer. I. The thermocouple. *Agr. Meteorol.* 3:433–447.

Peck, A. J., and R. M. Rabbidge. 1969. Design and performance of an osmotic tensiometer for measuring capillary potential. *Soil Sci. Soc. Am. Proc.* 33:196–202.

Penman, H. L. 1940. Meteorological and soil factors affecting evaporation from fallow soil. *Quart. J. R. Meteorol. Soc.* 66:401–410.

Penman, H. L. 1948a. Natural evaporation from open water, bare soil, and grass. *Proc. R. Soc. London*, A193:120–145.

Penman, H. L. 1948b. Physics in agriculture. *Sci. Instrum.* 25:425–432.

Penman, H. L. 1949a. The dependence of transpiration on weather and soil conditions. *J. Soil Sci.* 1:74–89.

Penman, H. L. 1949b. A general survey of meteorology in agriculture and an account of the physics of irrigation control. *Quart J. R. Meteoro. Soc.* 75:293–302.

Penman, H. L. 1951. The role of vegetation in meteorology, soil mechanics and hydrology. *Brit. J. Appl. Phys.* 2:145–151.

Peters, Doyle B., Robert M. Hagan, and Geoffrey B. Bodman. 1953. Available moisture capacities of soils as affected by addition of polyelectrolyte soil conditioners. *Soil Sci.* 75:467–471.

Philip, J. R. 1954a. Some recent advances in hydrologic physics. *J. Instrum. Eng. Aust.* 26:255–259.

Philip, J. R. 1954b. An infiltration equation with physical significance. *Soil Sci.* 77:153–157.

Philip, J. R. 1955. Numerical solution of equations of the diffusion type with diffusivity concentration-dependent. No. 391, *Trans. Faraday Soc.* 51:885–892.

Philip, J. R. 1957. Theory of infiltration: 1. The infiltration equation and its solution. *Soil Sci.* 83:345–357.

Philip, J. R. 1964. Similarity hypothesis for capillary hysteresis in porous materials. *J. Geophys. Res.* 69:1553–1562.

Philip, J. R. 1974a. Fifty years progress in soil physics. *Geoderma* 12:265–280.

Philip, J. R. 1974b. Recent progress in the solution of nonlinear diffusion equations. *Soil Sci.* 117:257–264.

Philip, J. R. 1977. Water on the earth. In: A. K. McIntyre (ed.), *Water, Plants and People.* Canberra Symposium, Australian Acad. Sci., Canberra.

Philip, J. R. 1980. Field heterogeneity: some basic issues. *Water Resour. Res.* 16:443–448.

Philip, J. R. 1983. Infiltration in one, two and three dimensions. In: *Proc. Nat. Conf. on Advances in Infiltration*, Chicago. Publ. 11-83, Am. Soc. Agr. Eng.

Philip, J. R., and D. A. De Vries. 1957. Moisture movement in porous materials under temperature gradients. *Trans. Am. Geophys. Union* 38:222–232.

Pierre, W. H., D. Kirkham, J. Pesek, and R. Shaw (eds.). 1966. *Plant Environment and Efficient Water Use*. Proc. of a 1965 Symposium at Iowa State University. Soil Sci. Soc. Am. and Am. Soc. Agron., Madison, WI.

Poiseuille, J. L. 1840–1841. Recherches experimentales sur le mouvement des liquides dans les tubes de tres petits diametres. *Compt. Rend.* 11:961–967, 1041–1048; 12:112–115.

Poulovassilis, A. 1962. Hysteresis of pore water, an application of the concept of independent domains. *Soil Sci.* 93:405–412.

Poulovassilis, A. 1970. Hysteresis of pore water in granular porous bodies. *Soil Sci.* 109:5–12.

Poulovassilis, A., and E. C. Childs. 1971. The hysteresis of pore water: the non-independence of domains. *Soil Sci.* 112:301–312.

Poulovassilis, A., and E. Tzimas. 1974. The hysteresis in the relationship between hydraulic conductivity and suction. *Soil Sci.* 117:250–256.

Puri, A. N. 1925. A critical study of the hygroscopic coefficient of soil. *J. Agr. Sci.* 15:272–283.

Puri, A. N. 1936. Soil physics. *Ann. Rev. Biochem. Allied Res. India* 7:146–150.

Puri, A. N. 1939. Physical characteristics of soils: V. The capillary tube hypothesis of soil moisture. *Soil Sci.* 48:505–520.

Puri, A. N. 1949. *Soils, Their Physics and Chemistry*. Reinhold, New York, 550 pp.

Puri, A. N., E. M. Crowther, and B. A. Keen. 1925. The relation between the vapour pressure and water content of soils. *J. Agr. Sci.* 15:68–88.

Quastel, J. H. 1952. Influence of organic matter on aeration and structure of soil. *Soil Sci.* 73:419–426.

Quirk, J. P. 1950. The measurement of stability of soil microaggregates in water. *Austr. J. Agr. Res.* 1:276–284.

Quirk, J. P. 1978. Some physico-chemical aspects of soil structure stability—a review. In: W. W. Emerson, R. D. Bond, and A. R. Dexter (eds.), *Modification of Soil Structure*. John Wiley, New York.???

Quirk, J. P., and R. K. Schofield. 1955. The effect of electrolyte concentration on soil permeability. *J. Soil Sci.* 6:163–178.

Raats, P. A. C. 1970. Steady infiltration from line sources and furrows. *Soil Sci. Soc. Am. Proc.* 34:709–714.

Raats, P. A. C. 1973. Unstable wetting fronts in uniform and non-uniform soils. *Soil Sci. Soc. Am. Proc.* 37:681–685.

Ramamoorthy, B. 1972, In memoriam, Dr. A. N. Puri. *Soil Sci.* 114:163.

Raney, W. A. 1949. Oxygen diffusion as a criterion of soil aeration. *Soil Sci. Soc. Am. Proc.* 14:61–65.

Rasmussen, W. D. 1960. *Readings in the History of American Agriculture*. Univ. Illinois Press, Urbana.

Rawlins, S. L. 1966. Theory for thermocouple psychrometers used to measure water potential in soil and plant samples. *Agr. Meteorol.* 3:293–310.

Rawlins, S. L. 1971. Some new methods for measuring the components of water potential. *Soil Sci.* 112:8–16.

Rawlins, S. L. 1973. Principles of managing high, frequency irrigation. *Soil Sci. Soc. Am. Proc.* 37:626–629.

Rawlins, S. L. 1976. Measurement of water content and state of water in soils. In: T. T. Kozlowski (ed.), *Water Deficits and Plant Growth*, Vol. IV, *Soil Water Measurement, Plant Responses, and Breeding for Drought Resistance.* Academic Press, New York, pp. 1–55.

Rawlins, S. L., and F. N. Dalton. 1967. Psychrometric measurement of soil water potential without precise temperature control. *Soil Sci. Soc. Am. Proc.* 31:297–301.

Rawlins, S. L., and W. H. Gardner. 1969. A test of the validity of the diffusion equation for unsaturated flow of soil water. *Soil Sci. Soc. Am. Proc.* 27:507–511.

Rawlins, S. L., and P. A. C. Raats. 1975. Prospects for high-frequency irrigation. *Science* 188:604–610.

Reeve, R. C. 1965. Modulus of rupture. In: C. A. Black, D. D. Evans, J. L. White, L. G. Ensminger, and F. E. Clark (eds.), *Methods of Soil Analysis.* Am. Soc. Agron Madison, WI, pp. 466–471.

Reynolds, O. 1883. An experimental investigation of the circumstances which determine whether the motion of the water shall be direct or sinuous, and of the law of resistance in parallel channels. *Proc. R. Soc. London* 35:84–99.

Richards, L. A. 1928. The usefulness of capillary potential to soil moisture and plant investigators. *J. Agr. Res.* 37:719–742.

Richards, L. A. 1931. Capillary conduction of liquids through porous mediums. *Physics* 1:318–333.

Richards, L. A. 1941. A pressure-membrane extraction apparatus for soil solution. *Soil Sci.* 51:377–386.

Richards, L. A. 1942. Soil moisture tensiometer materials and construction. *Soil Sci.* 53:241–248.

Richards, L. A. 1948. Porous plate apparatus for measuring moisture retention and transmission by soil. *Soil Sci.* 66:105–110.

Richards, L. A. 1949. Methods of measuring soil moisture. *Soil Sci.* 68:95–112.

Richards, L. A. 1950. Laws of soil moisture. *Trans. Am. Geophys. Union* 31:750–756.

Richards, L. A. 1953. Modulus of rupture as an index of crusting of soil. *Soil Sci. Soc. Am. Proc.* 17:321–323.

Richards, L. A. 1955. Water content changes following the wetting of bare soil in the field. *Soil Sci. Fla. Proc.* 15:142–148.

Richards, L. A. 1960. Advances in soil physics. *Trans. 7th Int. Congr. Soil Sci.* 1:67–79.

Richards, L. A., and Milton Fireman. 1943. Pressure plate apparatus for measuring moisture sorption and tansmission by soils. *Soil Sci.* 56:395–404.

Richards, L. A., and W. Gardner. 1936. Tensiometers for measuring the capillary tension of soil water. *J. Am. Soc. Agron.* 28:352–358.

Richards, L. A., and D. C. Moore. 1952. Influence of capillary conductivity and depth of wetting on moisture retention in soil. *Trans. Am. Geophys. Union* 33:531–540.

Richards, L. A., and G. Ogata. 1958. A thermocouple for vapor pressure measurements in biological materials and soil systems at high humidity. *Science* 128:1089–1090.

Richards, L. A., and G. Ogata. 1961. Psychrometric measurements of soil samples equilibrated on pressure membranes. *Soil Sci. Soc. Am. Proc.* 15:456–459.
Richards, L. A., and C. H. Wadleigh. 1952. Soil water and plant growth. In: B. T. Shaw (ed.), *Soil Physical Conditions and Plant Growth*, Vol. II. Agronomy Monographs, Am. Soc. Agron., Madison, WI, pp. 73–251.
Richards, L. A., and L. R. Weaver. 1943. Fifteen-atmosphere-percentage as related to the permanent wilting percentage. *Soil Sci.* 56:331–339.
Richards, L. A., W. R. Gardner, and G. Ogata. 1956. Physical processes determining water loss from soil. *Soil Sci. Soc. Am. Proc.* 20:311–314.
Richards, S. J. 1938. Soil moisture content calculations from capillary tension records. *Soil Sci. Soc. Am. Proc.* 3:57–64.
Richards, S. J. 1965. Soil suction measurements with tensiometers. In: C. A. Black, D. D. Evans, J. L. White, L. G. Ensminger, and F. E. Clark (eds.), *Methods of Soil Analysis*. Am. Soc. Agron., Madison, WI, pp. 153–163.
Richards, S. J., and A. W. Marsh. 1961. Irrigation based on soil suction measurements. *Soil Sci. Soc. Am. Proc.* 25:65–69.
Richards, S. J., and L. V. Weeks. 1953. Capillary conductivity values from moisture yield and tension measurement on soil columns. *Soil Sci. Soc. Am. Proc.* 17:206–209.
Richards, S. J., R. M. Hagan, and T. M. McCalla. 1952. Soil temperature and plant growth. In: B. T. Shaw (ed.), *Soil Physical Conditions and Plant Growth*. Am. Soc. Agron., Madison, WI, pp. 303–460.
Richtmyer, F. K. 1928 (1934 ed.). *Introduction to Modern Physics*. McGraw-Hill, New York.
Rijtema, P. E. 1959. Calculation of capillary conductivity from pressure plate outflow data with non-negligible memebrane impedance. *Netherlands J. Agr. Sci.* 7:209–215.
Rijtema, P. W., and H. Wassink. 1969. Water in the Unsaturated Zone, Procedings of the Wageningen Symposium. UNESCO/Int. Assoc. Scientific Hydrol., Paris.
Robins, J. R. 1952. Some thermodynamic properties of soil moisture. *Soil Sci.* 74:127–139.
Robinson, G. W. 1922. A new method for the mechanical analysis of soils and other dispersions. *J. Agr. Sci.* 12:287–291.
Robinson, G. W. 1924. On the form of mechanical composition curves of soils and other granular substances. *J. Agr. Sci.* 14:626–633.
Robinson, G. W. 1932. *Soils: Their Origin, Constitution and Classification*. T. Murby and Co., London, 300 pp.
Rode, A. A. 1947. (Transl. from Russian, 1961.) *The Soil Forming Process and Evolution*. Israel Program for Scientific Translation, Jerusalem (NSF, USDA, Washington, D.C.).
Rode, A. A. 1955. (Transl. from Russian, 1962.) *Soil Science*. Israel Program for Scientific Translation, Jerusalem (NSF, USDA, Washington, D.C.).
Rode, A. A. 1956. The water regimes of soils. (Transl. from Russian). *Pochvovedenie* 4:1–23.
Rodewald, H. 1902. Theorie der Hygroskopicitat. *Landw. Jehrb.*, 31:675–696.
Rodgers, A. D., III. 1949. *Liberty Hyde Bailey*. Princeton Univ. Press, Princeton, NJ.
Rose, C. W. 1968. Evaporation from bare soil under high radiation conditions. *Trans. 9th. Int. Cong. Soil Sci.* 1:57–66.

Rose, C. W., W. R. Stern and E. Drummond. 1965. Determination of hydraulic conductivity as a function of depth and water content for soil in situ. *Austr. J. Soil Res.* 3:1–9

Ruehrwein, R. A., and D. W. War. 1952. Mechanism of clay aggregation by polyelectrolytes. *Soil Sci.* 73:485–492.

Russell, E. J. 1912. *Soil Conditions and Plant Growth* (8th ed., 1950, recast and rewritten by E. Walter Russell.). Longmans, Green, London.

Russell, E. J. 1957. *The World of the Soil.* Collins, London, 237 pp.

Russell, E. J. 1962. The rebirth of soil science in Great Britain. (On the ninetieth birthday of Sir E. John Russell). *Soil Sci.* 94:199–213.

Russell, E. W. 1934. The interaction of clay with water and organic liquids as measured by specific volume changes and its relation to their phenomena of crumb formation in soils. *Phil. Trans. R. Soc. London*, 233A:361–389.

Russell, E. W. 1938a. Physical basis of soil structure. *Sci. Prog.* 32:660–676.

Russell, E. W. 1938b. *Soil Structure.* Sci. Tech. Comm. 37, Imp. Bur. Soils, Harpenden, 40 pp.

Russell, J. C. 1946. The movement of water in soil columns and the theory of the control sections. *Soil Sci. Soc. Am.* 11:119–123.

Russell, J. C. and W. W. Burr. 1925. Studies on the moisture equivalent of soils. *Soil Sci.* 19:251–266.

Russell, M. B. 1943. The utility of the energy concept of soil moisture. *Soil Sci. Soc. Am. Proc.* 7:90–94.

Russell, M. B. 1949. Methods of measuring soil structure and aeration. *Soil Sci.* 68:25–35.

Russell, M. B. 1952. Aeration and plant growth. In: B. T. Shaw (ed.), *Soil Physical Conditions and Plant Growth*, Vol. II. Agronomy Monographs, Am. Soc. Agron., Madison, WI, pp. 73–251.

Russell, M. B. 1959. Plant responses to differences in soil moisture. *Soil Sci.* 88:179–183.

Russell, M. B. 1960. Some physical aspects of plant growth. *Soil Sci. Soc. Am. Proc.* 24:439–440.

Russell, M. B., and L. A. Richards. 1938. Heat of wetting of soils. *Proc. Iowa Acad. Sci.* 45:179–185.

Schamp, N., J. Huylebroeck, and M. Sadones. 1975. Adhesion and absorption phenomena in soil conditioning. In: W. C. Moldenhauer, W. R. Gardner, C. E. Clapp, M. M. Mortland, W. H. Gardner, and C. I. Rich (eds.), *Soil Conditioners.* Publ. 7, Soil Sci. Soc. Am., Madison, WI, pp. 13–23.

Schloesing, A. T. 1872. Influence du terreau sur l'ameublissement des sols. *Comp. Rend.* 74:1408–1411.

Schloesing, A. T. 1874. Sur la constitution des argiks (duexieme note), Sur la constitution des argiles:kaolins. *Compt. Rend.* 79:376–380, 473–477.

Schofield, R. K. 1935. The pF of the water in soil. *Trans. 3rd Int. Cong. Soil Sci.* 2:37–48.

Schofield, R. K. 1943. The role of moisture in soil mechanics. *Chem. Indust.* 62:339–341.

Schofield, R. K. 1946. Ionic forces in thick films of liquid between charged surfaces. *Trans. Faraday Soc.* 42B:219–225.

Schofield, R. K. 1950. Soil moisture and evaporation. *Trans. 4th Intr. Congress Soil Sci.* 2:20–28.

Schofield, R. H., and H. L. Penman. 1949. The principles governing transpiration by vegetation. *Inst. Civ. Eng. Proc. Conf. Biol. Civ. Engng.*, 1948, pp. 75–84.
Schroeder, D. 1980. Soil science in the federal republic of Germany—past, present, future. *Soil Sci.* 130:178–179.
Schubler, G. 1830. *Grundsatze der Agricultur-Chemie.* In Baumgartners buchhandlung, Liepzig.
Schumacher, W. 1864. *Die Physik. I. Die Physik des Bodens.* Wiegandt and Hempel, Berlin.
Sharma, M. L., Goro Uehara, and J. A. Mann, Jr. 1969. Thermodynamic properties of water adsorbed in dry soil surfaces. *Soil Sci.* 107:86–93.
Shaw, B. T. 1942. The nature of colloidal clay as revealed by the electron microscope. *J. Phys. Chem.* 46:1032–1043.
Shaw, B. T. (ed.). 1952. *Soil Physical Conditions and Plant Growth*, Vol. II. Agronomy Monographs, Am. Soc. Agron. Academic Press, New York.
Shaw, B. T., and L. D. Baver. 1939. Heat conduction as an index of soil moisture. *J. Am. Soc. Agron.* 31:886–889.
Shaw, B. T., and L. D. Baver. 1940. An electrothermal method for following moisture changes of soil *in situ. Soil Sci. Soc. Am. Proc.* 4:78–83.
Shaw, C. F. 1926. The effect of a paper mulch on soil temperature. *Hilgardia* 1:341–364.
Shaw, C. F., and A. Smith. 1927. Maximum height of capillary rise starting with soil at capillary saturation. *Hilgardia* 2:400–409.
Slater, C. S. 1948. The flow of water through soil. *Agr. Eng.* 29:119–124.
Slatyer, R. O. 1967. *Plant-Water Relationships.* Academic Press, New York.
Slichter, C. S. 1898. Theoretical Investigation of the Motion of Ground Waters. 19th Ann. Rept. Pt. 2, U.S. Geol. Surv., U.S. Govt. Printing Office, Washington, D.C. pp. 295–384.
Slichter, C. S. 1966. *Science in a Tavern, Essays and Diversons on Science in the Making.* Univ. Wisconsin Press, Madison, WI, 206 pp.
Smiles, D. E. 1974. Infiltration into a swelling material. *Soil Sci.* 117:140–147.
Smith, W. O. 1933. Minimum capillary rise in an ideal soil. *Physics* 4:184–193.
Smith, W. O. 1936. Sorption in an ideal soil. *Soil Sci.* 41:209–230.
Smith, W. O. 1939. Thermal conductivities in moist soils. *Soil Sci. Soc. Am. Proc.* 4:32–40.
Soane, B. D. 1967. Dual energy gamma-ray transmission for coincident measurement of water content and dry bulk density of soil. *Nature* (London) 214:1273–1274.
Soane, B. D. 1968. A gamma-ray transmission method for the measurement of soil density in field tillage studies. *J. Agric. Eng. Res*, 13(4):340–349.
Spanner, D. C. 1951. The Peltier effect and its use in the measurement of suction pressure. *J. Exp. Bot.* 2:145–68.
Staple, W. J., and J. J. Lehane. 1954. Movement of Water in unsaturated soils. *Can. J. Agr. Sci.* 34:329–342.
Stokes, Sir G. G. 1898. Stream-line Motion of a Viscous Film. II. Mathematical Proof of the Identity of the Stream Lines Obtained by Means of a Viscous Film with Those of a Perfect Fluid Moving in Two Dimensions. Report of the 68th Meeting of the British Assoc. Adv. Sci., Bristol. John Murray, London, pp. 143–144.

Swartzendruber, D. 1966. Soil-water as described by transport coefficients and functions. *Adv. Agron.* 189:327–370.

Swartzendruber, D. 1969. The flow of water in unsaturated soil. In: R. J. M. De Wiest (ed.), *Flow in Porous Media*, Academic Press, New York, pp. 53–57.

Swartzendruber, D. 1974. Infiltration of constant-flux rainfall into soil as analyzed by the approach of Green and Ampt. *Soil Sci.* 117:272–281.

Swartzendruber, D. 1977. Soil physics--reflections and perspectives. In: M. D. Thorne (ed.) *Agronomists and Food: Contributions and Challenges*. Spec. Publ. 30, Am. Soc. Agron., Madison, WI.

Swartzendruber, D. 1978. In recognition of Don Kirkham on his seventieth birthday. *Soil Sci.* 125:65–67.

Tanner, C. B. 1957. Factors affecting evaporation from plants and soils. *J. Soil Water Cons.* 12:221–227.

Tanner, C. B. 1964. Franklin Hirum King (1948–1911). Mimeograph, Dept. Soils. Univ. Wisconsin, Madison, WI.

Tanner, C. B. 1967. Measurement of evaporation. In: R. M. Hagan, H. R. Haise, and T. W. Edminster (eds.), *Irrigation of Agricultural Lands*. Am. Soc. Agron., Madison, WI, pp. 534–555.

Tanner, C. B. 1968. Evaporation of water from plants and soils. In: T. T. Kozlowskis (ed.), *Water Stress and Plant Growth*, Vol. I. Academic Press, New York, pp. 74–104.

Tanner, C. B., and D. E. Elrick. 1958. Volumetric porous pressure plate apparatus for moisture hysteresis measurements. *Soil Sci.Soc. Am. Proc.* 22:575–576.

Taylor, G. S. and W. P. Martin. 1953. Effect of soil-aggregating chemicals on soils. *Agr. Eng.* 34:550–554.

Taylor, H. M. 1971. Soil conditions as they affect plant establishment, root development and yield. In: K. K. Barnes, W. M. Carleton, and R. I. Throckmorton (eds.), *Compaction of Agricultural Soils*. Am. Soc. Agr. Eng., St. Joseph, MI, pp. 292–305.

Taylor, H. M., and Herbert R. Gardner. 1963. Penetration of cotton seedling taproots as influenced by bulk density, moisture content, and strength of soil. *Soil Sci.* 96:153–156.

Taylor, S. A. 1950. Oxygen diffusion in porous media as a measure of soil aeration. *Soil Sci. Soc. Am. Proc* 14:55–61.

Taylor, S. A. 1952. Estimating the integrated soil moisture tension in the root zone of growing plants. *Soil Sci.* 73:331–339.

Taylor, S. A. 1958. The activity of water in soils. *Soil Sci.* 86:83–90.

Taylor, S. A. 1965. Willard Gardner, 1883–1964. *Soil Sci.* 100:79–82.

Taylor, S. A., and G. L. Ashcroft. 1972. *Physical Edaphology*. Freeman Press, San Francisco.

Taylor, S. A., and J. W. Cary, 1964. Linear equations for the simultaneous flow of matter and energy in a continuous soil system. *Soil Sci. Soc. Am. Proc.* 28:167–172.

Taylor, S. A., and G. L. Stewart. 1960. Thermodynamic properties of soil water. *Soil Sci. Soc. Am. Proc.* 24:243–247.

Taylor, S. A., D. D. Evans, and W. D. Kemper. 1961. Evaluating Soil Water. Bull. 426, Utah Agr. Exp. Sta., Logan, UT, 67 pp.

Terzaghi, K., and R. B. Peck. 1948. *Soil Mechanics in Engineering Practice*. John Wiley and Sons, New York.

Thomas, M. D. 1921. Aqueous vapor pressure of soils. *Soil Sci.* 11:409–434.
Thomas, M. D. 1924. Aqueous vapor pressure of soils. II. Studies in dry soils. *Soil Sci.* 17:1–18.
Thorne, M. D. and M. B. Russell. 1948. Dielectric properties of soil moisture and their measurement. *Soil Sci. Soc. Am. Proc.* 12:66–72.
Thornwaite, W. W., and B. Holzman. 1942. Measurement of Evaporation from Land and Water Surfaces. Tech. Bull. 817, USDA, Washington, D.C.
Topp, G. C. 1971. Soil-water hysteresis: the domain theory extended to pore interaction conditions. *Soil Sci. Soc. Am. Proc.*, 35:219–225.
Topp, G. C., and E. E. Miller. 1966. Hysteretic moisture characteristics and hydraulic conductivities for glass-bead media. *Soil Sci. Soc. Am. Proc.* 30:156–162.
Tull, J. 1731. *The Horse Hoeing Husbandry*. Jethro Tull, London.
U. S. Salinity Laboratory Staff (L. A. Richards, ed). 1954. *Diagnosis and Improvement of Saline and Alkali Soils.* Handbook 60, U.S. Salinity Lab., Riveside, CA, 160 pp.
Vachaud, G., and J. L. Thony. 1971. Hysteresis during infiltration and redistribution in a soil column at different initial water contents. *Water Resour. Res.* 7:111–127.
van Bavel, C. H. M. 1952a. Evapotranspiration estimates as criteria for determining time of irrigation. *Agr. Eng.* 33:417–e418.
van Bavel, C. H. M. 1952b. Gaseous diffusion and porosity in porous media. *Soil Sci.* 73:91–104.
van Bavel, C. H. M. 1959. Soil densitometry by gamma transmission. *Soil Sci.* 87:50–58.
van Bavel, C. H. M., N. Underwood, and S. R. Ragar. 1957. Transmission of gamma radiation by soils and soil densitometry. *Soil Sci. Soc. Am. Proc.*, 21:588–591.
van Bemmelen, J. M. 1878–1879. Das Absorptionsvermogen der Ackererde. *Lanw. Vers. Stat.* 21:135–1191; 23:265–303.
van Bemmelen, J. M. 1910. *Die Absorption*. Ostwald, Dresden.
van Wijk, W. R., and D. A. De Vries. 1954. Evapotranspirataion. *Netherlands J. Agr. Sci.* 2:105–119.
van Wijk, W. R., (ed.). 1963. *Physics of Plant Environment.* North Holland Publ. Amsterdam pp. 382.
Veihmeyer, F. J., and A. H. Hendrickson. 1931. The moisture equivalent as a measure of field capacity. *Soil Sci.* 32:181–193.
Veihmeyer, F. J., and A. H. Hendrickson. 1933. Some Plant and Soil-Moisture Relations. Bull. 15, Am. Soil Surv. Assoc.
Veihmeyer, F. J., N. E. Edlefsen, and A. H. Hendrickson. 1943. Use of tensiometers in measuring availability of water to plants. *Plant Physiol.* 18:66–78.
Vershinin, P. V. 1958 (Engl. Transl. 1971). *The Background of Soil Structure.* Israel Program for Scientific Transl., Jerusalem.
Visralingham, M., and J. D. Tandy. 1972. The neutron method for measuring soil moisture content. A review. *J. Soil Sci.* 23:499–511.
Vomocil, J. A. 1954. *In situ* measurement of soil bulk density. *Agr. Eng.* 35:651–654.
Vomocil, J. A., E. R. Fountain, and R. J. Reginato. 1958. Effect of moisture content on tensile strength of unsaturated glass bead systems. *Soil Sci. Soc. Am. Proc.* 26:409–412.

Vomocil, J. A., L. J. Waldron, and W. J. Chancellor. 1961. Soil tensile strength by centrifugation. *Soil Sci. Soc. Am. Proc.* 25:176–180.
von Sachs, Julius. 1859. Bericht uber die physiologische Thatigkeit und der Versuchsstation in Tharandt. *Landwirthschaftlichen Versuchs-Stationen*, 1:235 (cited by Briggs and Shantz, 1912).
Wadleigh, C. H., and H. G. Gaugh. 1948. Rate of leaf elongation as affected by the intensity of the total soil moisture stress. *Plant Physiol.* 23:485–495.
Wadsworth, H. A. 1931. Further observations upon the nature of capillary rise through soils. *Soil Sci.* 32:417–434.
Wang, H. F. 1986. Charles Sumner Slichter—An Engineer in Mathematician's Clothing. In: E. R. L. and S. Ince (eds.), *History of Hydrology*. Am. Geophys. Union, Washington, D.C. (In press).
Warington, R. 1900. *Lectures on Some of the Physical Properties of Soil*. Clarendon Press, Oxford, England.
Warkentin, B. P., G. H. Bolt, and R. D. Miller. 1957. Swelling pressure of montmorillonite. *Soil Sci. Soc. Am. Proc.* 21:495–497.
Warrick, A. W., J. H., Kichen, and J. L. Thames. 1972. Solutions for miscible displacement of soil water with time-dependent velocity and dispersion coefficients. *Soil Sci. Soc. Am. Proc.*, 36:863–867.
Washburn, E. W. 1921. The dynamics of capillary flow. *Phys. Rev.* Ser. 2, 17:273–283.
Watson, K. K. 1966. An instantaneous profile method for determining the hydraulic conductivity of unsaturated porous materials. *Water Resour. Res.* 2:709–715.
Weeks, L. E., and W. G. Colter. 1952. Effect of synthetic soil conditioners on erosion control. *Soil Sci.* 73:473–484.
Whitney, M. 1892. Some Physical Properties of Soils in Their Relation to Moisture and Crop Distribution. Bull. 4, U.S. Weather Bur., Washington, D.C.
Whitney, M. 1909. Soils of the Untied States. Bull. 55, U.S. Bur. Soils, Washington, D.C.
Whitney, M., and F. K. Cameron. 1903. The chemistry of the soil as related to crop production. Bull. 22, U.S. Bur. Soils, Washington, D.C.
Widstoe, J. A. 1911. *Dry Farming.* The Macmillan Co., New York, 445 pp.
Widstoe, J. A. 1914. *The Principles of Irrigation Practice.* The Macmillan Co., New York, 496 pp.
Widstoe, J. A., and W. W. McLaughlin. 1902. Irrigation Experiments in 1901 (on the college farm). bull. 80, Utah Agr. Exp. Sta., Logan, UT.
Wiebe, H. H., R. W. Brown, T. W. Daniel, and E. Campbell. 1970. Water potential measurements in trees. *Bioscience* 20:225–226.
Wiegand, C. L., and S. A. Taylor. 1962. Temperature depression and temperature distribution in drying soil columns. *Soil Sci.* 94:75–79.
Wilkinson, G. E., and A. Klute. 1962. The temperature effect on the equilibrium energy status of water held by porous media. *Soil Sci. Soc. Am. Proc.* 17:326–329.
Winterkorn, H. F. 1943. The condition of water in porous systems. *Soil Sci.* 56:109–115.
Wollny, M. E. 1878–1881. Untersuchungen uber den Einfluss der Farbe des

Bodens auf dessen Erwarmung, Erste und Zweite Mitteilung. *Forsch. Geb. Agrik.-Phys.* 1:45–69; 4:327–365.

Wollny, M. E. 1884–1885. Untersuchungen uber die Kapillare Leitung des Wassers im Boden. *Forsch. Geb. Agrik. Phys.* 7:269–308; 8:206–220.

Yoder, R. E. 1936. A direct method of aggregate analysis of soils and a study of the physical nature of erosion losses. *J. Am. Soc. Agron.* 28:337–351.

Youngs, E. G. 1957. Moisture profiles during vertical infiltration. *Soil Sci.* 84:283–290.

Youngs, E. G. 1958. Redistribution of moisture in porous materials after infiltration. *Soil Sci.* 86:117–125, 202–207.

Youngs, E. G. 1964. An infiltration method of measuring the hydraulic conductivity of unsaturated porous materials. *Soil Sci.* 97:307–311.

Youngs, E. G. 1974. Water-table heights in homogenous soils drained by nonideal drains. *Soil Sci.* 117:295–300.

Youngs, E. G., and A. J. Peck. 1964. Moisture profile development and air compression during water uptake by bounded propus bodies: 1. Theoretical introduction. *Soil Sci.* 98:290–294.

Youngs, E. G., G. D. Towner, and A. Poulovassilis. 1974. In memoriam, Ernest Carr Childs. *Soil Sci* 117:241–242.

Zaslavsky, D. 1964. Saturated and unsaturated flow equation in an unstable porous medium. *Soil Sci.* 98:317–321.

Zaslavsky, D., and I. Ravina. 1965. Measurement and evaluation of hydraulic conductivity through the moisture movement method. *Soil Sci.* 100:104–108.

Zingg, A. W. 1951. Evaluation of the erodibility of field surfaces with a portable wind tunnel. *Soil Sci. Soc. Am. Proc.* 15:11–17.

Zunker, F. 1930. Das Verhalten des Bodens sum Wasser. *Handbuch der Bodenlehre* 6:66–220.

Other References

Amerman, C. R. 1973. Hydrology and Soil Science. In: R. R. Bruce, K. W. Flad, and H. M. Taylor, *et al.* (eds.). *Field Soil Water Regime.* pp. 167–180.

Bear, Firman E. 1961. Soil-plant research in the United Kingdom. *Soil Sci.* 92:1–77.

Bird, R. B., W. E. Stewart, and E. N. Lightfoot. 1960. *Transport Phenomena.* John Wiley and Sons, New York, 780 pp.

Gardner, W. H. 1986. Growth of Understanding of the Physics of Soil and Water, the Earth's Most Critical Resources—A Historical Perspective. Presidential Address, Spec. Publ. No. 4, Pacific Division, Am. Assoc. Adv. Sci., San Francisco.

Gill, W. R., and G. E. Vanden Berg. 1967. *Soil Dynamics in Tillage and Traction.* ARS, USDA, U.S. Govt. Printing Office, Washington, D.C.

Gulhati, N. D., and W. C. Smith. 1967. Irrigated agriculture: an historical review. In: R. M. Hagan, H. R. Haise, and T. W. Edminster (eds.), *Irrigation of Agricultural Lands.* Am. Soc. Agron., Madison, WI, pp. 3–9.

Hall, Sir A. D. 1909. *Fertilizers and Manures* (Revised 4th ed., 1947, by A. M. Smith). Wyman and Sons, Ltd., London.

Hanks, R. J., and G. L. Ashcroft. 1980. *Applied Soil Physics: Soil Water and Temperature Applications.* Springer-Verlag, Berlin.

Marshall, T. J., and J. W. Holmes. 1979. *Soil Physics*. Cambridge Univ. Press, Cambridge.

Meinzer, O. E. 1942. *Hydrology*. McGraw-Hill, New York, 703 pp.

Monteith, J. L. 1973. *Principles of Environmental Physics*. American Elsevier, New York.

Muskat, M. 1937 (1939 ed.). *Physical Principles of Oil Production*. McGraw-Hill, New York.

Muskat, M. 1946. *The Flow of Homogeneous Fluids Through Porous Media*. J. W. Edwards, Ann Arbor, MI.

Olmstead, L. B., and W. O. Smith. 1938. Water relations of soils. In: C. E. Kellog (ed.) *Fundamentals of Soil Science*, Part IV, *Soils and Men*. U. S. Govt. Printing Office, Washington, D.C., pp. 897–910.

Rose, C. W. 1966. *Agricultural Physics*. Pergamon Press, Sydney.

Salmon, S. C., And A. A. Hanson. 1964. *The Principles and Practice of Agricultural Research*. Leonard Hill, London.

Scheidegger, A. E. 1957 (1960 ed.). *The Physics of Flow Through Porous Media*. The Macmillan Co., New York, 313 pp.

Schroeder, D. 1980. Soil science in the federal republic of Germany–past, present, future. *Soil Sci*. 130:178–179.

Slatyer, R. O. 1967. *Plant–water relationships*. Academic Press, New York, London.

Stefferund, A. (ed.). 1957. *Soil, the Yearbook of Agriculture*. U.S. Govt. Printing Office, Washington, D.C.

Van Wijk, W. R. (ed.). 1963. *Physics of Plant Environment*. North Holland Publ., Amsterdam, 382 pp.

Nitrous Oxide Emission from Soils

K. L. Sahrawat and D. R. Keeney*

I. Introduction

Nitrous oxide (N_2O) constitutes only 350 parts per billion (ppb) of the atmosphere. Its production by biochemical processes in soils has long been recognized (Delwiche, 1981; Payne, 1981a, b). However, recent

*Department of Soil Science, University of Wisconsin-Madison, Madison, WI 53706, USA.
Contribution from the college of Agricultural and Life Sciences and by the National Science Foundation.
Present address (KLS): International Crops Research Institute for the Semi-Arid Tropics (ICRISAT), Patancheru P.O., Andhra Pradesh 502324, India.

© 1986 by Springer-Verlag New York Inc.

concern about the involvement of N_2O in promoting destruction of the stratospheric ozone layer, which protects the biosphere from harmful ultraviolet radiation (Crutzen, 1970, 1971, 1974, 1981; Johnston, 1971, 1977; CAST, 1976; McElroy *et al.*, 1977; Liu *et al.*, 1977), and also the concern that the increased N_2O may contribute to the "greenhouse effect" caused by increased CO_2 concentration (Yung *et al.*, 1976), have resulted in extensive research on how N_2O emissions are affected by fertilizer N usage for meeting food, feed, and fiber requirements of the world (Delwich, 1981; Payne, 1981b; Keeney, 1982; Firestone, 1982; Freney and Simpson, 1983).

Nitrous oxide is the source of various oxides of N that catalyze the destruction of ozone by photochemical reactions (Crutzen, 1981). These can be represented as follows:

$$O_3 + h\nu \rightarrow O + O_2$$
$$O + NO_2 \rightarrow NO + O_2$$
$$NO + O_3 \rightarrow NO_2 + O_2$$

This series of reactions results in the net loss of ozone, which can be represented as:

$$2\ O_3 \rightarrow 3\ O_2$$

Measurement of N_2O emissions from soils as a result of use of N in inorganic or organic forms is of interest to atmospheric scientists, agronomists, and soil scientists. Atmospheric scientists' concern arises from the effect of N_2O on ozone destruction, while agronomists and soil scientists view N_2O emission as a mechanism of N loss from soils. Additional interest in N_2O measurement as a measure of direct denitrification loss of N has arisen from a recent finding that acetylene blocks the conversion of N_2O to N_2 during denitrification in soils (Balderston *et al.*, 1976; Yoshinari and Knowles, 1976; Yoshinari *et al.*, 1977; M. S. Smith *et al.*, 1978). Although there are problems associated with the use of acetylene to measure N loss through denitrification, this finding has stimulated interest in N_2O emission as a direct measurement of in situ denitrification under field conditions (Tiedje, 1982).

Several recent reviews focus on the processes leading to N_2O emission from soils (Trudinger *et al.*, 1980; Delwiche, 1981; Payne, 1981b; Firestone, 1982). The purpose of this review is to put together the available information on biological and chemical processes in soils that are connected with N_2O production from different soil ecosystems, mechanism of N_2O emission under managed and native soil ecosystems, and the management factors potentially of use in control of N_2O emission in relation to processes contributing to these emissions.

II. Processes Contributing to N_2O Production

While denitrification has been considered the major source of N_2O in soils and waters (CAST, 1976; Delwiche, 1981; Payne, 1981a), it is now recognized that nitrification may also contribute to N_2O emissions (Bremner and Blackmer, 1981). Nitrous oxide also can be formed by chemical reactions when $[NO_2]^-$ or NH_2OH are decomposed in acid soils, producing small amounts of N_2O (Corbet, 1935; Arnold, 1954; Nelson and Bremner, 1970; Bremner and Blackmer, 1980; Nelson, 1982), but these processes likely contribute relatively little to N_2O producton in soils as compared to nitrification and denitrification.

A. Nitrification and N_2O Emission

The first indication of N_2O formation during nitrification was obtained by Corbet (1935) with cultures of nitrifiers. He suggested that N_2O was produced by the microorgganisms that oxidized $[NH_4]^+$ or NH_2OH to $[NO_2]^-$. Later studies (Yoshida and Alexander, 1970, 1971) showed that the cells of *Nitrosomonas europaea* produced N_2O during oxidation of $[NH_4]^+$ or NH_2OH to $[NO_2]^-$. Ritchie and Nicholas (1972, 1974) provided evidence that $[NO_2]^-$ was reduced to N_2O by nitrate reductase synthesized by *Nitrosomonas europaea*. It was generally believed that much of the N_2O produced in soils was the result of denitrification of $[NO_3]^-$ under anaerobic or microanaerobic conditions. However, recent studies (Blackmer and Bremner, 1977; Bremner and Blackmer, 1978, 1979; Freney *et al*., 1978, 1979; Goodroad and Keeney, 1984a, 1985; Robertson and Tiedje, 1984; Sahrawat *et al*., 1985) have shown that N_2O is produced in soils under aerobic conditions in the presence of nitrifiable forms of N and may also represent a major source of marine N_2O (Elkins *et al*., 1978). Blackmer *et al*. (1980) found that N_2O was evolved when sterilized soils were incubated under aerobic conditions after treatment with cultures of nitrifying microorganisms by *Nitrosomonas, Nitrosospira*, or *Nitrosolobus* species and $(NH_4)_2SO_4$ (Table 1). Nitrous oxide was not evolved when the sterilized soils, treated with $[NH_4]^+$, were not inoculated with nitrifying microorganisms, indicating that the nitrifiers were capable of producing N_2O in soils.

Bremner and Blackmer (1981), in a summary of the recent literature on N_2O emission associated with nitrification, concluded that N_2O was evolved from well-aerated soils, and that N_2O production was related to the nitrifiable N content and was not affected by the addition of $[NO_3]^-{-}N$. Emissions of N_2O from well-aerated soils treated with $[NH_4]^+{-}N$ or urea were greatly retarded by the nitrification inhibitor, nitrapyrin [2-chloro-g-(tri-chloromethyl)pyridine], which retards nitrification of $[NH_4^+]$-N.

Field studies of N_2O emission from Iowa soils showed that the amounts of N_2O evolved from plots treated with $(NH_4)_2SO_4$ or urea markedly exceeded those from plots receiving the same amount of N as $Ca(NO_3)_2$. The fertilizer-induced emissions of $N_2O{-}N$ observed during 96 days from plots treated with $(NH_4)_2SO_4$ (125 to 250 kg N ha $^{-1}$) constituted only 0.11 to 0,18% of the fertilizer N applied. The amount of

Table 1. Effects of ammonia-oxidizing chemoautotriphic nitrifiers on emission of N_2O from sterilized soil treated with ammonium sulfate[a,b]

Nitrifier strain used	N_2O-N evolved in 21 days (ng g^{-1} of soil)
None	< 1
Nitrosomonas europaea	712
Nitrosomonas Tara	30
Nitrosomonas E-K	473
Nitrosolobus AV-3	102
Nitrosolobus Bearden	126
Nitrosopira AV-2	208

[a]From Blackmer *et al.*, (1980) with permission.

[b]Soil samples (30 g) of sterilized storden soil were treated with $(NH_4)_2SO_4$ (100 mg kg^{-1} soil) and inoculated with nitrifier specified and incubated (30°C; 0.23 ml water per gram of soil) under air in 1.2-L flasks. Flask atmospheres were renewed after 2, 7, and 14 days.

N_2O emitted from plots receiving $Ca(NO_3)_2$ did not increase with the amount of N applied and was not appreciably greater than the amounts emitted from control plots (Table 2, Breitenbeck *et al.*, 1980). Bremner *et al.* (1981) studied the effect of anhydrous ammonia (NH_3) fertilization of three Iowa field soils (250 kg N ha^{-1}) on N_2O emissions and found that the emissions of N_2O—N from the fertilized soils in 139 days averaged 15.6 kg ha^{-1} and ranged from 12.1 to 19.6 kg ha^{-1}. The emissions of N_2O—N from the unfertilized soils averaged 2.0 kg ha^{-1}. The fertilizer-induced N_2O emissions represented 4.0 to 6.8% of the fertilizer N applied (Table 3).

Table 2. Effects of different forms and amounts of fertilizer nitrogen on N_2O evolved from field plots during 96 days[a]

Nitrogen fertilizer			
Form	Amount (kg N ha^{-1})	Amount of N_2O evolved (g N ha^{-1})	N_2O-N evolved as % of N added
None	—	330	—
$Ca(NO_3)_2$	125	382	0.04
$Ca(NO_3)_2$	250	358	0.01
Urea	125	503	0.14
Urea	250	617	0.12
$(NH_4)_2SO_4$	125	557	0.18
$(NH_4)_2SO_4$	250	613	0.11

[a]From Breitenbeck *et al.*, *Geophys. Res. Lett.* 7, 85–88, 1980; copyright by the American Geophysical Union.

Table 3. Effect of anhydrous ammonia fertilization on N_2O emission from soils[a]

Soil			Amount of N_2O-N evolved			
			96 days		139 days	
Series	Subgroup	Fertilizer treatment[b]	kg ha^{-1}	% of N applied	kg ha^{-1}	% of N applied
Harps	Typic Calcia-quolls	None	1.6	—	1.7	—
		AA	14.9	5.3	15.0	5.3
Webster	Typic Hapla-quolls	None	2.3	—	2.5	—
		AA	19.2	6.8	19.6	6.8
Canisteo	Typic Hapla-quolls	None	1.9	—	2.0	—
		AA	12.0	4.0	12.0	4.0
Mean		None	2.0	—	2.0	—
		AA	15.4	5.4	15.6	5.4

[a](Adapted from Bremner *et al., J. Environ. Quality* 10, 1981, 77–80 by permission of the Am. So Agro., Crop Sci. Soc. Am., Soil Sci. Soc. Am.)

[b]AA, anhydrous ammonia was applied at the rate of 250 kg N ha^{-1}.

Nitrous oxide emissions exceeded considerably those in similar field studies using $Ca(NO_3)_2$, urea, or $(NH_4)_2SO_4$ (for comparison, see results in Tables 2 and 3).

Freney *et al.* (1979) found that N_2O was emitted from soils at water contents ranging from air dry to field capacity. The rate of emission increased with increase in temperature from 4 to 37°C. Nitrous oxide formation resulted from microbiological processes and was prevented by autoclaving and inhibited by $HgCl_2$ or toluene. Apparently N_2O was formed during nitrification at lower moisture content (see also Bremner and Blackmer, 1978, 1979).

Cochran *et al.* (1981) followed N_2O emission via nitrification from a fallow field fertilized with 0, 55, 110, and 220 kg N ha^{-1} as NH_3. The daily fluxes of N_2O ranged from 1 to 15 g N ha^{-1} day^{-1}, with the least from unfertilized plots and maximum from the plots fertilized with 200 kg N. However, even at the highest rate of N application less than 0.1% of the applied N was recovered as N_2O after 5 weeks. Nitrate tended to accumulate in soil fertilized with higher rates of NH_3.

B. Factors Affecting N_2O Production via Nitrification

Soil and environmental factors that affect nitrification also influence N_2O emissions via the nitrification process. Soil pH, organic matter, water and oxygen status, and the capacity of soils to supply nitrifiable N ($[NO_3]^{-1}$ production) have greatest influence on N_2O production in soils.

Among the environmental factors, temperature, rainfall (soil water content), and the soil ecosystem and nutrient status affect N_2O emissions.

1. Nitrifier Culture Studies

Nitrous oxide emissions from the cultures of nitrifiers isolated from soils are also influenced by cell density, age of cells, pH, O_2 concentration, temperature, concentration of such elements as phosphate, and the concentration of substrate ($[NH_4]^+$ or NH_2OH). Yoshida and Alexander (1970, 1971) found that N_2O emissions by *Nitrosomonas europaea* increased with increasing concentration of substrate ($[NH_4]^+$ or NH_2OH), high pH, or an increase in temperature from 0 to 35°C. Cell age and density and phosphate concentration also greatly influenced N_20 evolution. They found that, in addition to *Nitrosomonas* species, several heterotrophic microorganisms, such as *Bacillus subtilis, Escherichia coli, Aerobacter aerogenes, Aspergillus flavus*, and *Penillium atrovenetum*, were capable of forming N_2O from $[NO_3]^-$ or $[NO_2]^-$.

Goreau *et al.* (1980), in a study with *Nitrosomonas* species isolated from marine environments, found that low O_2 partial pressures (0.005 to 0.2 atm or 0.18 to 7 mg L^{-1}) decreased the rate of $[NO_2]^-$ production from 3.6 $\times 10^{-10}$ to 0.5×10^{-10} mmol of $[NO_2]^-$ per cell but increased emissions of N_2O from 1×10^{-12} to 4.3×10^{-12} mmol N day^{-1} per cell. Nitrifying bacteria from the genera *Nitrosomonas, Nitrosospira, Nitrosolobus*, and *Nitrosococcus* emitted similar quantities of N_2O at 0.2 atm O_2. Nitrite-oxidizing bacteria (*Nitrobacter* species) did not emit measurable quantities of N_2O during growth. Cates (1983) studied the physiological and N substrate factors affecting N_2O-N production by cultures of nitrifers maintained at pH 7.5 and at 23°C under aerobic conditions (0.2 atm O_2) and reported that the physiological growth stage of the nitrifiers greatly influenced N_2O production. The emission of N_2O-N ranged from 0.2×10^{-3} to 23×10^{-3} mol per mole of $[NH_4]^+$ oxidized. Nitrate level did not affect N_2O emission rates.

Nitrous oxide can be produced by microorganisms other than classical nitrifiers or denitrifiers. Several *Rhizobium* spp. were shown by Zablotowic *et al.* (1978) to be capable of reducing $[NO_3]^-$ to N_2O under anoxic conditions, and O'Hara *et al.* (1983) confirmed that N_2O was the dominant gaseous N product. O'Hara *et al.* (1983) found the denitrifying enzymes of *Rhizobium japonicum, R. hupini*, and *R. meliloti* to be O_2 sensitive. *Rhizobium*-dependent denitrification may be an important avenue of N loss in certain ecological situations (Daniel *et al.*, 1980; O'Hara *et al.*, 1984). M. S. Smith and Zimmerman (1981) found 209 of 214 soil bacterial isolates capable of producing N_2O from $[NO_3]^-$, but only 46 of these were true respiratory denitrifiers. They were unable to evaluate the significance of these nondenitrifying $[NO_3]^-$ reducers as a source of soil N_2O but felt they might contribute to N_2O evolution from soil.

Similarly, Bleakley and Tiedje (1982) and Burth *et al.* (1982) have isolated numerous heterotrophic bacteria, yeasts, and fungi capable of producing N_2O. Therefore, it is apparent that N_2O can be produced by nonnitrifiers and nondenitrifiers and that N_2O production involves a complex ecological niche in the soil N cycle.

2. Soil Studies

Bremner and Blackmer (1981) have summarized the factors affecting N_2O emissions during nitrification. Nitrous oxide was emitted from soils under very well-aerated conditions and N_2O emission also was observed from soil samples containing less than 5% water. Nitrous oxide emission rates increased with increase in pH and soil organic matter content and were strongly increased on amendment of soils with nitrifiable sources of inorganic N or plant and organic residues. Increase in soil temperature from 5 to 30°C increased the rate of emission of N_2O from well-aerated soil samples. The temperature effect was far more pronounced when the samples were amended with 200 μg g^{-1} of $[NH_4]^+-N$. Nitrous oxide emissions from unamended samples increased from 2 to 5 ng g^{-1} soil with increase in temperature from 5 to 30°C and the $[NH_4]^+$-amended samples had N_2O-N emission increases of from 16 to 242 ng g^{-1} soil in 16 days (Bremner and Blackmer, 1981). Similarly, Freney *et al.* (1979) reported that N_2O emission rates from Australian soils increased with increase in soil temperature from 4 to 37°C. Blackmer *et al.* (1982) found that N_2O emissions from field, moist unamended soil increased with increase in temperature from 10 to 40°C and Q_{10} values, the rate of reaction increase for each 10° increase in temperature, varied from 1.35 to 2.28 (Table 4).

Soil water content also greatly affects N_2O production in soils under aerobic conditions. Freney *et al.* (1979) reported that N_2O emissions

Table 4. Effect of temperature on emission of N_2O from soil[a,b]

Temperature (°C)	N_2O-N evolved in 21 hr (ng g^{-1} soil) (s.d.)	Q_{10}
10	8.3 (0.5)	—
20	18.9 (0.8)	2.28
30	37.6 (0.5)	1.99
40	50.9 (3.3)	1.35

[a](Adapted from Blackmer *et al.*, *Soil Sci. Soc. Am. J.*, 46, 1982, 937–942, by permission of Soil Sci. Soc. Am.).

[b]Samples of field-moist soil (Typic Hapla-quoll, 23% water) were incubated at the specified temperature for 21 hr.

increased with increase in soil water from air dry to field capacity, and they concluded that production of N_2O was continuous at all moisture contents. Studies reported by Bremner and Blackmer (1981) also indicated that emissions of N_2O increased with increase in soil moisture from 20 to 150% of the water-holding capacity (WHC). It was further observed that emissions of N_2O during 30 days from soils amended with $[NH_4]^+$ exceeded those from samples amended with $[NO_3]^-$ when the amended soils were saturated with water to promote denitrification. They also found that air drying of soil samples prior to aerobic incubation greatly increased N_2O emission rates.

The emissions of N_2O from soils are also greatly affected by addition of chemicals that retard nitrification. For example, nitrapyrin, $HgCl_2$, toluene, and CS_2 have been reported to retard N_2O emissions associated with nitrification (Freney *et al.*, 1978, 1979; Blackmer *et al.*, 1980; Bremner and Blackmer, 1980, 1981). Acetylene, which is known to inhibit oxidation of $[NH_4]^+$ by nitrifying microorganisms (Hynes and Knowles, 1978; Walter *et al.*, 1979), is also reported to retard N_2O emissions associated with nitrification in well-aerated soils (Bremner and Blackmer, 1979, 1981; Blackmer *et al.*, 1980; Aulakh *et al.*, 1984b).

Goodroad and Keeney (1984a) investigated N_2O production in aerobically incubated soils as affected by pH, temperature, and water content. Samples of a silt loam collected from plots maintained at pH 4.7, 5.1, or 6.7 were used. It was found that nitrification of added $[NH_4]^+-N$ increased with increase in soil pH (4.7 to 6.7), temperature (10 to 30°C), and water content (10 to 30% v/v). The rate of N_2O emission was proportional to $[NO_3]^-$ production and on the average 0.1% of the nitrified N was evolved as N_2O-N in 5 days. The ratio of N_2O-N to $[NO_3]^--N$ produced was greatly affected by soil pH and temperature, particularly at 5 kPa soil water potential (Table 5). Addition of nutrients, such as P, and liming materials, such as $CaCO_3$, can also affect N_2O evolution from soils in some situations. For example, Minami and Fukushi (1983) found that application of P or $CaCO_3$ to four Japanese soils increased N_2O emission under aerobic incubation. However, P-induced N_2O emissions were larger than those obtained with $CaCO_3$. The addition of $CaCO_3$ and P enhanced the growth of ammonium oxidizers and accumulation of $[NO_2]^--N$ which probably accounted for enhanced N_2O evolution.

Addition of $CaCO_3$ to six acid forest soils increased emissions of N_2O during 28 days of aerobic incubation at 30°C under field-moist conditions, but P application had little effect on N_2O emissions (Table 6, Sahrawat *et al.*, 1985). It was found that $CaCO_3$ addition greatly enhanced ammonification or non- or slowly nitrifying soils and both ammonification and nitrification in nitrifying soils. From 0.03 to 0.3% of the $[NH_4]^+-N$ nitrified was released as N_2O-N. Calcium carbonate addition increased the soil pH to the neutral range (6.7 to 7.5) and increased

Table 5. Ratio of N_2O-N to $[NO_3]^-$-N produced in Plano silt loam soils of varying pH at different water contents and temperatures under aerobic incubation for 5 days[a,b]

Soil water	$(N_2O\text{-}N/[NO_3]^-\text{-}N) \times 10^{-3}$								
	pH 4.7			pH 5.1			pH 6.7		
(m^3/m^{-3})	10°C	20°C	30°C	10°C	20°C	30°C	10°C	20°C	30°C
0.1	1.5	1.5	1.9	0.7	0.9	0.9	1.0	0.9	1.0
0.2	1.1	2.0	2.4	0.9	0.8	0.9	0.8	1.0	1.0
0.3	1.3	1.9	3.3	0.8	1.2	11.5	1.0	6.0	5.7

[a]Reprinted with permission from *Soil Biol. Biochem.*, 16 L. L. Goodroad and D. R. Keeney, pp. 39–43; copyright 1984 by Pergamon Press, Ltd.

[b]Soil samples were treated with 100 mg N as $(NH_4)_2SO_4$ kg^{-1} soil. Soil water potential (kPa) is 1600, 20 and 5; water-filled porosity is 18, 36, and 54% at water content of 0.1, 0.2, and 0.3 m^3 m^{-3}, respectively.

Table 6. Effect of phosphorus and $CaCO_3$ on N_2O-N production during aerobic incubation of six forest soils after 28 days at 30°C[a,b]

Site		Forest floor type	pH ($CaCl_2$)	Organic matter (%)	N_2O-N (μg N kg^{-1} dry soil)		
Soil	Stand[c]				Control	+P	+$CaCO_3$
Histosol	Hemlock	Mor	3.9	59	7	3	4
Alfisol	Sugar maple	Mull	5.1	22	19	29	29
Alfisol	Red oak	Mor	4.3	16	49	48	110
Alfisol	White oak	Mor	4.4	25	36	28	109
Spodosol	White pine	Mor	4.4	42	56	62	96
Entisol	Red pine	Disturbed	4.4	2.4	6	4	25

[a]Reprinted from Sahrawat *et al.* (1985) by permission of *Forest Sci.*

[b]Field-moist samples (10 g) from Blackhawk Island, Wisconsin (Pastor *et al.*, 1984) were treated with 1 ml of water, 1 ml of water containing 1000 mg L^{-1}P (as KH_2PO_4), or 0.2 g (2%) of powdered $CaCO_3$.

[c]Hemlock, *Tsuga canadensis* (L.) Carr.; sugar maple, *Acer saccharum* Marsh; red oak, white oak, *Quercus rubra* L., *Q. alba* L.; red pine, white pine, *Pinus resinosa* Ait., *P. strobus* L.

nitrification and N_2O emissions. Adding P and $CaCO_3$ together did not affect N transformations, including N_2O evolution, compared to $CaCO_3$ alone.

Keeney *et al.* (1985) showed that CO_2 concentration in the soil atmosphere can affect N_2O evolution associated with nitrification and denitrification. Some results of this study are shown in Table 7. It was found that an increase in CO_2 concentration from 0.3 to 100% increasingly retarded the rate of nitrification in soil. Nitrous oxide production associated with nitrification increased as CO_2 increased from 0.3 to 2.6% and tended to be greater as CO_2 concentration increased to 73%. At 100% CO_2, no N_2O was produced during 7 days of incubation at 25°C. Carbon dioxide, however, did not affect N_2O production or reduction in the saturated, $[NO_3]^-$-amended soil nor the rate of N_2O reduction in anaerobic environments. It was concluded that increased CO_2 in arable soils containing large

Table 7. Effect of varying concentrations of CO_2 on N_2O production and on the ratio of N_2O produced to NO_3 produced in soil under aerobic conditions in 7 days[a,b]

Treatment					
CO_2 (%)	$[NH_4]^+$ (μg N g^{-1})	$[NO_3]^-$ (μg N g^{-1})	CO_2 7 days (%)	N_2O-N (ng N g^{-1})	N_2O-N to $[NO_3]^-$-N ($\times 10^{-3}$)
Air	0	0	3.3	117 ± 2	3.4
22	0	0	25	83 ± 14	4.4
50	0	0	49	154 ± 11	22
73	0	0	70	544 ± 55	181
100	0	0	100	0	—
Air	0	95	3.1	123 ± 10	3.6
23	0	95	25	168 ± 23	7
50	0	95	47	250 ± 28	25
73	0	95	71	716 ± 303	90
100	0	95	100	46,400 ± 1350	—
Air	90	0	3.6	396 ± 126	6
2.6	90	0	6.2	1,120 ± 208	19
5.2	90	0	8.6	972 ± 88	16
11	90	0	14	1,616 ± 637	32
23	90	0	26	1,111 ± 104	28
50	90	0	49	1,034 ± 152	74
73	90	0	71	1,270 ± 562	423
100	90	0	100	0	—

[a]Reprinted with permission from *Soil Biol. Biochem.*, 17, D. R. Keeney *et al.*, 571–573; copyright 1985, Pergamon Press, Ltd.

[b]Plano silt loam (10 g) was incubated at 25°C with 2.5 ml of water or with 2.5 ml of a solution containing either 900 μg NO_3-N (as KNO_3) or 950 μg $[NH_4]^+$-N $(NH_4)_2SO_4$ under the specified CO_2 concentrations.

amounts of nitrifiable N could result in increased N_2O production during nitrification. The effect was not large until CO_2 concentration was greater than 50% of the soil atmosphere.

C. Denitrification and N_2O Emission

According to the *Glossary of Soil Science* (Soil Science Society of America, 1979), denitrification is "the microbial reduction of $[NO_3]^-$ or $[NO_2]^-$ to gaseous N as molecular N (N_2) or as an oxide of N" (ps). Several reviews are available (Garcia, 1975a; Delwiche and Bryan, 1976; Focht and Verstraete, 1977; Payne, 1973, 198b; Delwiche, 1981; Knowles, 1981; Firestone, 1982). Additionally, various aspects of N_2O production via denitrification have been covered by several recent reviews (Crutzen, 1981; Ingraham, 1981; Bryan, 1981; Schroeder, 1981; Rolston, 1981; Hahn, 1981; Payne, 1981a, b).

1. Laboratory Studies

Beijernick and Minkman (1910) and Suzuki (1912) reported N_2O to be one of the gases evolved by bacteria in anaerobic cultures and soils. Wijler and Delwiche (1954) studied the gaseous products of denitrification in soils using ^{15}N-tagged KNO_3. They identified evolution of N_2O, N_2 and NO gases during dentrification. Nitrous oxide was the major production of dentrification and the ratio of N_2O/ to N_2 evolved was affected by soil pH. Above pH 7, N_2O was readily reduced to N_2. However, at pH < 7, the rate of reduction of N_2O to N_2 declined. Evolution of and the relative proportions of N_2O and N_2 were not affected by the applications of organic residues. There was no evidence of any nonenzymatic denitrifications (Van Slyke reaction) taking place. At low soil pH (4.9) N0 was produced in amounts equal to that of N_2O evolved. Chemical decomposition of $[NO_2]^-$ formed from $[NO_3]^-$ was suggested as possible pathway for NO formation (also see Cady and Bartholomew, 1961, 1963):

$$3\ HNO_2 \rightleftharpoons 2\ NO + HNO_3 + H_2O$$

Subsequent studies have found that numerous facultative heterotrophs can produce N_2O from $[NO_3]^-$ and can further reduce N_2O to N_2 (Delwiche, 1959). Payne (1981a, b) has discussed the intermediates formed during denitrification. It is now certain that denitrification involves several enzymatic and chemical reactions and follows the sequence:

$$2\ [NO_3]^- \rightarrow 2[NO_2]^- \rightarrow 2\ NO \rightarrow N_2O \rightarrow N_2$$

Studies with cultures of various denitrifying bacteria (Payne, 1981a, b) indicate that 13 genera of include species capable of reducing $[NO_3]^-$ to N_2 via $[NO_2]^-$, NO, and N_2O. Nitrous oxide production by denitrifying microorganisms in cultures have been reported by many workers (Najjar and Allen, 1954; Delwiche, 1959; Baalsrud and Baalsrud, 1954; Barbaree and Payne, 1967; Miyata, 1971; Ishaque and Aleem, 1973; Pichinoty *et al.*, 1978; M. S. Smith and Tiedje, 1980; Payne, 1981a, b). Many studies of denitrification in soils have identified N_2O as one of the intermediates, and the ratios of N_2 to N_2O evolved vary greatly (Arnold, 1954; Wijler and Delwiche, 1954; Nommik, 1956; Cooper and Smith, 1963; McGarity and Hauck, 1969; Stefanson, 1973; Burford and Bremner, 1975; Rolston *et al.*, 1976; Letey *et al.*, 1980a, b; Payne 1981a, b).

2. Field Studies

Field studies of N_2O evolution as a mechanism of fertilizer N loss have been summarized by Rolston (1981). The amount of N_2O evolved varied from nil to as high as 8 kg N ha^{-1} day^{-1}. The highest emissions were reported in summer from experimental plots maintained in a wet state and treated with manure (Table 8). Much research indicates a range in N_2O emission ranging from <0.01 to 1 kg N_2O-N ha^{-1} day^{-1} (Burford and Millington, 1968; Burford and Stefanson, 1973; Dowdell and Smith, 1974; Rolston *et al.*, 1976, 1978; Freney *et al.*, 1978; Focht, 1978; McKenney *et al.*, 1978; Ryden *et al.*, 1978, 1979; Findlay and McKenney, 1979; Hutchinson and Mosier 1979, 1981; Ryden, 1981, 1983; C. J. Smith *et al.*, 1981a; Burford *et al.*, 1981; Webster and Dowdell, 1982; Mosier *et al.*, 1982, 1983; Aulakh *et al.*, 1982, 1983a, b; Armstrong, 1983; Goodroad and Keeney, 1984b; Goodroad *et al.*, 1984) depending on N fertilization rate, source of N applied, soil water content, tillage practices, and the prevailing soil temperature.

The fluxes of N_2O reported from organic soils are higher than those reported for mineral soils. For example, Terry *et al.* (1981) reported that N_2O-N fluxes from cultivated organic soils of south Florida ranged from 0.004 kg N ha^{-1} day^{-1} during the dry periods to 4.5 kg N ha^{-1} day^{-1} following rainfall events. The total N_2O-N emissions during 1 yr (1979 to 1980) amounted to 165, 97, and 48 kg N ha^{-1} for the fallow, grass cropped, and sugarcane (*Saccharum officinarum* L.) cropped organic soils, respectively (Table 9). Similarly, Duxbury *et al.* (1982) found that the maximum daily N_2O-N fluxes from cropped mineral soils ranged from 0.19 to 0.63 kg N ha^{-1} during a 2-yr study compared to 2.9 to 4.6 kg N ha^{-1} for an organic soil during the same period. The annual emission of N_2O-N from the mineral soils ranged from 0.9 to 0.42 kg N ha^{-1}, whereas for the organic soils they ranged from 7 to 165 kg N ha^{-1}. Goodroad and Keeney (1984b) also found high N_2O emission rates from drained organic soils.

D. Factors Affecting N_2O Production via Denitrification

Factors that affect denitrifications in soils also affect N_2O emissions but the ratio of N_2 to N_2O evolved is sensitive to environmental and soil

Table 8. Denitrification loss of nitrogen in summer as N_2O and N_2 from plots maintained at different soil-water pressure heads[a,b]

Treatment[c]	Soil water pressure head (cm)	Surface flux: N_2O (kg N ha^{-1})	Surface flux: N_2 (kg N ha^{-1})	Flux at 60-cm depth[d], $N_2 + N_2O$ (kg N ha^{-1})	Total (kg N ha^{-1})	Total (% of fertilizer N)
Manure	−15	9.9	198	10	218	73
Manure	−70	5.4	42	0	47	16
Cropped	−15	4.3	30	5.5	40	14
Cropped	−70	1.8	7	0	9	3
Uncropped	−15	2.1	5.7	2.5	10	3
Uncropped	−70	0.6	3.6	0	4	1

[a]Adapted from Rolston *et al.*, *Soil Sci. Soc. Am. J.* 42, 1978, 863–869, by permission of Soil Sci. Soc. Am.

[b]The plots were established on a Yolo loam (Typic Xerotthents) soil. The average soil temperature at the 5-cm depth during July (when the study was done) was approximately 23°C.

[c]All treatments received 300 kg N ha^{-1} as KNO_3.

[d]Calculated from the $^{15}N_2$ and N_2O concentration gradients between the 30- and 60-cm depths and from the average gaseous diffusion coefficient between two depths of 5 cm^2 hr^{-1} corresponding to a soil air content of 0.12 cm^3 cm^{-3}.

Table 9. Effects of cropping on nitrous oxide emissions from Pahokee Muck during 1979 and 1980[a]

	Treatment					
	Fallow		Grass		Sugarcane	
Period of Study	Range (kg N_2O-N ha^{-1})	Total (kg N_2O-N ha^{-1})	Range (kg N_2O-N ha^{-1})	Total (kg N_2O-N ha^{-1})	Range (kg N_2O-N ha^{-1})	Total (kg N_2O-N ha^{-1})
May 1979 to April 1980	0.4–60.4	164.9	0.4–58.8	96.8	0.3–26.1	48.4

From Terry *et al.* (1981); reprinted with permission of *APCA J.*

Table 10. Summary of factors that affect the relative proportion of N_2O to N_2 evolved via the denitrification process[a]

Factor	Influence on N_2O/N_2 ratio
Nitrate concentration	Ratio increases with increasing $[NO_3]^-$
Nitrite concentration	Ratio increases with increasing $[NO_2]^-$
Oxygen concentration	Ratio increases with increasing O_2
pH	Ratio increases with pH decrease; pH also enhances the effect of $[NO_3]^-$
Sulfide	Ratio increases wtih increasing sulfide
Carbon	Ratio generally decreases with increasing C availability
Redox potential (Eh)	Ratio not affected by Eh changes below 0 mV
Enzyme status	Ratio can be increased or decreased by synthesis of or absence of N_2O reducing activity relative to preceding reductases

[a]From Firestone (1982). Reproduced from Nitrogen in Agricultural Agronomy, Agronomy Monograph No. 22, 1982, pages 289–326. By permission of the American Society of Agronomy.

factors (Table 10). Rolston (1981), Payne (1981b), and Firestone (1982) have discussed factors affecting denitrification and N_2O production in soils. Mosier *et al.* (1983) focused on factors that affected N_2O evolution from cropped and native soils with a view to modeling N_2O emissions in soil.

1. Substrate Concentration and Availability

Presence of NO_3 or $[NO_2]^-$ is a prerequisite for N_2O production via denitrification, In a recent review, Mosier *et al.* (1983) analyzed the data obtained from a number of field studies and found that N_2O production via denitrification was correlated with the $[NO_3]^-$ content of soil ($r = 0.65$, $n = 136$). They found that N_2O fluxes could be represented by first-order kinetics when the $[NO_3]^-$ content of the soils did not exceed 40 μg $[NO_3]^-{-}N\ g^{-1}$ soil. At higher $[NO_3]^-{-}N$ concentration reduction of N_2O to N_2 may be slowed (Blackmer and Brenmer, 1978). This conclusion is in accord with the earlier findings that at high $[NO_3]^-{-}N$ concentrations the dominant gasous product of denitrification is N_2O (Nommik, 1956; Cooper and Smith, 1963; Blackmer and Bremner, 1978; Cho and Mills, 1979; Rolston, 1981). However, the effect of $[NO_3]^-$ on reduction of N_2O has been found to be less in organic soils and sediments (Terry and Tate, 1980).

Evidence presented by Firestone *et al.* (1980) suggests that $[NO_3]^-$

concentration affects the evolution of N_2O and the ratio of N_2 to N_2O, although the total loss of N via denitrification in soils may be independent of $[NO_3]^-$ concentration over a wide range (Wijler and Delwiche, 1954; Nommik, 1956; Bremner and Shaw, 1958).

2. Soil Water Content and Soil Drying

The soil water regime affects denitrification and gasous products of denitrification directly and also indirectly by affecting the diffusion of O_2 into and through the soil. Nommik (1956) found that the denitrification rate was proportional to the water content of soils, and at the lower water contents the gaseous product of denitrification mainly consisted of N_2O. Molecular nitrogen evolution was favored by strictly anaerobic conditions. Studies by Blackmer and Bremmer (1976), Bremmer and Blackmer (1979), Freney *et al.* (1978), and Focht *et al.* (1979) suggest that N_2O initially produced when water content is greater than the field capacity is reduced to N_2, so that N_2O emission rates are lower than N_2 emission rates at field capacity. Mosier *et al.* (1983) reported that soil relative water content (RWC, an index of soil water content at wilting point divided by water content at field capacity minus the water content at wilting point) was related to N_2O emissions in field soils (r = 0.52, n = 136). Nitrous oxide fluxes were low at RWC $<$ 0.35 and increased as RWC increased.

Similarly, studies by Terry *et al.* (1981), Duxbury *et al.* (1982), and Goodroad and Keeney (1984b) found that N_2O fluxes from mineral and organic soils were many times lower during dry periods than shortly after rainfall events when the soils were saturated (Table 11).

Drying or freezing of soil samples prior to anaerobic incubation also increases the rate of denitrification and the evolution of N_2O and N_2 (e.g., McGarity, 1962; Myers and McGarity, 1971; McKenzie and Kurtz, 1976; Patten *et al*., 1980). Studies by Patten *et al.* (1980) showed that even partial drying of field-moist soils increased the emission of $(N_2O + N_2)-N$ from soils incubated anaerobically after treatment with $[NO_3]^-$. Air-dry storage of soils increased the amounts of $(N_2O + N_2)-N$ evolved when these soils were incubated anaerobically. It was suggested that drying or air-dry storage of soil increased the soil's capacity to denitrify and to evolve gaseous products of denitrification (N_2O and N_2) under anaerobic conditions by enhancing the amount of soil organic matter available to the denitrifiers. C. J. Smith and Patrick (1983) studied the effect of several anaerobic and aerobic cycles on N_2O emission from $(NH_4)_2SO_4$-amended soil suspensions and found that no N_2O was evolved from continuously anaerobic samples. Continuously aerobic samples produced small amounts of N_2O, whereas alternate anaerobic–aerobic cycles of varying duration increased N_2O emissions several fold (Table 12).

Table 11. Nitrous oxide emissions from cultivated mineral and organic soils during 2 yr of study[a]

Crop	N fertilizer or manure treatment[b]	Maximum observed daily N_2O kg ha^{-1}	Annual N_2O emissions	
			May 1979–May 1980 (kg ha^{-1})	May 1980–May 1981 (kg ha^{-1})
Mineral Soils				
Alfalfa	0	0.63	4.2	2.3
Field corn A	130(M)	0.22	2.4	2.9
	130(F)	0.28		2.2
Field corn B	130(M)	0.19		3.8
Timothy weed	0	0.23	1.7	0.9
Organic Soils				
Onions	170	4.5	85	72
Sweet corn	170	2.9	76	152
Sugarcane (Florida)	0	3.1	48	7
St. Augustine grass (Florida)	0	4.6	97	16
None (fallow, Florida)	0	4.5	165	59

[a]Reprinted by permission from Duxbury *et al.*, *Nature*, 298, 462–464, copyright (c) 1982 Macmillan Journals Limited.

[b]Field corn sites received 20 kg N ha^{-1} added as either urea–ammonium nitrate (1:1) fertilizer side dressed 40 days after planting (F) or manure (45 tons ha^{-1}) applied before planting (M). Fertilizer containing NH_4NO_3 was broadcast on the organic soil sites in New York before planting.

Table 12. Effects of continuous aerobic or anaerobic conditions and of alternate anaerobic and aerobic treatments on N_2O emission from soil suspension[a,b]

Treatment	N_2O-N evolved in 56 days (ng N g^{-1} dry soil)
Aerobic	790
Anaerobic	0
Alternate anaerobic/aerobic (7/7 days)	7,190
Alternate anaerobic/aerobic (14/14 days)	15,670

[a]Adapted with permission from *Soil Biol Biochem*., 15, C. J. Smith and W. H. Patrick, pp. 693–697; copyright 1983 Pergamon Press, Ltd.

[b]Soil samples (400 g) were amended with $[NH_4]^+$-N (92 mg N kg^{-1} dry soil) and ground rice straw (0.5%, w/w) and water was added to form soil: water ratio of 1:5 and pH of 6.2 ± 0.2.

3. Oxygen Supply

The absence of O_2 or its restricted availability in soils or at soil microsites is one prerequisite for denitrification (see Rolston, 1981; Payne, 1981a, b; Firestone, 1982, for reviews). However, the water content of soils also indirectly affects the oxygen content, and hence denitrification, as well as N_2O production by affecting the rate of diffusion of O_2 (Wijler and Delwiche, 1954; Nommik, 1956; Bremner and Shaw, 1958; Pilot and Patrick, 1972; Focht, 1974; Firestone and Tiedje, 1979; Rolston, 1981; Letey *et al*., 1980a, b; Firestone, 1982; Cho, 1982); K. A. Smith (1980) has developed an excellent model of the extent of anaerobic zones in soil aggregates.

Focht (1974) used a zero-order kinetic model to estimate the production of N_2O and N_2 in soils and stated that aeration affected the ratio of N_2 to N_2 evolved from soils. The rates of denitrification increase with decreased aeration, but the N_2O to N_2 ratio declines because of greater rate of reduction of N_2O to N_2 under anaerobic conditions (also see Letey *et al*., 1981).

Firestone *et al*. (1980) studied the effects of factors controlling the production of N_2O through biological denitrification by bacteria cultures and soils and found that increased oxygen content enhanced production of N_2O relative to N_2 during denitrification. Under anaerobic conditions, N_2O production initially was also found to increase, but this was followed by N_2O consumption in the system and its conversion to N_2 by nitrous oxide reductase (Letey *et al*., 1981). C. J. Smith *et al*. (1983) reported that the critical redox potential (Eh) for N_2O reduction and production occurred at +250 mV for pH 6, 7, and 8.5 and at +300 mV for pH 5 for the soil-water systems. They, as well as Letey *et al.* (1981), have confirmed the theory of Freney *et al.* (1979), that the soil can act as a N_2O sink under anoxic conditions. Letey *et al.* (1981) found that N_2O emissions were higher in soils with fluctuating redox potential established by wetting and drying cycles.

4. Carbon Supply

It has long been recognized that availability of C enhances the loss of $[NO_3]^-$ or $[NO_2]^-$ through denitrification (Rolston, 1981; Payne, 1981a, b; Firestone, 1982). Studies by Wijler and Delwiche (1954), Nommik (1956), and M. S. Smith and Tiedje (1979) showed that C supply decreased the ratio of N_2O to N_2 evolved during denitrification in soils.

The total fluxes of N_2O-N have been reported to be many times higher in organic soils with higher contents of soil organic matter and are also affected by cropping (Tables 11 and 12; Terry *et al.*, 1981; Duxbury *et al*, 1982). Burford and Bremmer (1975) and Patten *et al.* (1980) found that N_2O or N_2 production in a number of Iowa soils was positively correlated with total organic C, water-soluble C, and mineralizable C (as measured by CO_2 production). The highest correlations (r = 0.98 and 0.99) were reported between the amounts of $(N_2 + N_2O)-N$ evolved and the amounts of CO_2 produced during anaerobic incubation of soils treated with $[NO_3]^-$ (Patten *et al.*, 1980).

The availability of C as an electron doner influences N_2O production in soils but the relationship is not direct. Often, changes in C availability are associated with factors that may affect N_2O production (Firestone, 1982; Koskinen and Keeney, 1982). Carbon supplied by application of organic residues increases not only N_2O production in soils but also the rate of N_2O reduction to N_2 (Wijler and Delwiche, 1954; Nommik, 1956; Firestone *et al.*, 1980; Letey *et al.*, 1980a, b, 1981). The rhizosphere of a plant may be a better source of energy for denitrification (Garcia, 1975b; Rolston, 1981; Firestone, 1982). However, the effect of the rhizosphere on N_2O production is not established. It is speculated that N_2O production may be greater in the rhizosphere zone to start with until nitrous oxide reductase is synthesized or rerepressed by anoxic conditions. This should decrease N_2O fluxes and increase N_2 production.

The depth within the soil profile where denitrification occurs also affects the N_2O mole fraction evolved. If denitrification occurs very near to the soil surface, N_2O produced can readily diffuse into the atmosphere, whereas if denitrification occurs deeper in the profile there is more possibility that the N_2O produced will be further reduced to N_2. Thus, the mole fraction of N_2O $[(N_2O-N)/(N_2O + N_2)-N]$ evolved is larger when denitrification occurs very close to the surface and smaller when the site of denitrification is deeper in the soil profile (Rolston, 1981).

5. Temperature

Temperature greatly influences denitrification rates in soils (Bremner, 1978; Broadbent and Clark, 1965; Rolston, 1981; Knowles, 1981; Payne, 1981a, b; Firestone, 1982). In earlier studies on the effects of temperature on denitrification, Nommik (1956) and Bremner and Shaw (1958) indicated that temperature was very important in controlling the rate of denitrification in soils and the temperature optimum for denitrification was suggested to be between 60 and 70°C. Nommik (1956) also noted that amounts of gaseous products formed were greatly influenced by temperature. He showed that while the rate of (N_2O + N_2)−N evolved from anaerobic soils treated with $[NO_3]^-$ increased with the increase in temperature from 3 to 65°C, evolution of N_2O became an increasingly

Table 13. Amounts of N_2 and N_2O evolved from nitrate-amended soil incubated under He at different temperatures[a,b]

Temperature (°C)	Form of N (μg N per 50 g soil)	Time (hr) 24	48	96
7	N_2	26	42	86
	N_2O	98	145	209
15	N_2	49	84	163
	N_2O	71	127	195
25	N_2	88	337	664
	N_2O	255	361	299
40	N_2	145	295	1117
	N_2O	749	2247	2659
50	N_2	953	7816	7609
	N_2O	4274	158	0
60	N_2	8884	8726	8159
	N_2O	0	0	0
65	N_2	8850	8553	7739
	N_2O	0	0	0
67	N_2	8956	8142	8707
	N_2O	0	0	0
70	N_2	262	695	851
	N_2O	3084	4171	5786
75	No N_2 or N_2O evolved			

[a]Adapted from Keeney *et al. Soil Sci. Soc. Am. J.* 43, 1979, 1124–1128, by permission of Soil Sci. Soc. Am.

[b]Plano silt from (Typic Argiudoll) soil samples (50 g) were treated with 6 ml of water containing 5 mg $[NO_3]$-N [as $Ca(NO_3)_2$] and incubated under He after adjusting soil water content to 40% (w/w) at the specified temperatures.

smaller component of gaseous N evolved and also that N_2O persisted for shorter times with increasing temperature.

Focht (1974) and Focht and Verstraete (1977), on the other hand, suggested that temperature had little effect on the gaseous composition of denitrification products. Focht (1974) did not include temperature in the zero-order kinetic model suggested for studying N_2O and N_2 production through denitrification. Bailey (1976) found, as did Nommik (1956), that greater proportions of N_2O were evolved at lower temperatures (6 to 8°C, 15°C) than at higher (30°C).

Keeney *et al.* (1979) examined the relative proportion of N_2O and N_2 evolved from a $[NO_3]^-$-amended silt loam during anaerobic incubation (under He atmosphere) in the temperature range of 7 to 75°C (Table 13). They found that the denitrification rate increased with temperature above 15°C, with an apparent temperature optimum of 60 to 67°C. Nitrous oxide evolution in relation to total $(N_2O + N_2)-N$ declined with time but was still 44 to 50% of the gas produced at 16 days at 7 and 15°C. Nitrous oxide comprised up to 88% of the gas produced at 40°C and after 2 days. However, at 50°C, most of the gas was N_2O after 1 day but at 50 to 70°C, N_2O had disappeared by 1 day. Denitrification was found to be occurring at 70°C but at a reduced rate. Nitrous oxide represented more than 85% of the N evolved at 70°C. Denitrification did not occur at 75°C. It was suggested that the commonly stated temperature optima for denitrification might be too high for true biological denitrification because of the dominance of temperature-dependent chemical reactions at temperatures higher than 50°C.

As suggested by Bailey (1976) and Keeney *et al.* (1979), temperature effects on N_2O production in soils may be complicated by accumulation of $[NO_2]^-$, which could be involved in chemodenitrification. Also in response to temperature, differential organic matter mineralization may change the redox potential, which further complicates prediction of N_2O production (Keeney *et al.*, 1979).

6. pH

Soil pH is an important factor in controlling the rate of denitrification. It is generally accepted that the optimal pH range for denitrification is between 7.0 and 8.0. Although denitrification occurs in soils over a wide range of pH, it is often reported that the denitrification rate is slow below pH 6 (Broadbent and Clark, 1965; Delwiche and Bryan, 1976; Focht and Verstraete, 1977; Rolston, 1981; Payne, 1981a, b; Knowles, 1981; Delwiche, 1981; Firestone, 1982).

It is now accepted that as the soil pH declines below 6.0, the evolution of N_2O relative to N_2 increases (Rolston, 1981; Knowles, 1981; Payne, 1981a, b; Firestone, 1982). For example, Wijler and Delwiche (1954) and Nommik (1956) found that production of N_2O relative to N_2 was much higher in soils with pH lower than 6.0, and this was because reduction of

N_2O was inhibited at low pH. These results have been confirmed by subsequent research (Hauck and Melsted; Cady and Bartholomew, 1960; Matsubara and Mori, 1968; Focht, 1974; Firestone *et al.*, 1980; Koskinen and Keeney, 1982).

Firestone *et al.* (1980) found that pH 4.9 or 6.5 had little effect on the ratio of N_2O to N_2 evolved at very low levels of $[NO_3]^-$ (<0.5 μg g^{-1}) but that the pH effect was large at 10 μg g^{-1}, indicating an interaction between pH and $[NO_3]^-$ substrate levels on the gaseous products of denitrification.

Koskinen and Keeney (1982) studied the effect of soil pH on the rate and products of denitrification in a soil that had been maintained at various pH levels in the field for 18 yrs. This avoided additions of salts to manipulate soil pH, which might have caused perturbations in the native microflora. It was found that N_2O comprised 83% of the gaseous N products for pH 4.6 and 5.4 when 90% of the $[NO_3]^-$-N (100 μg g^{-1} of $[NO_3]^-$ added) had been denitrified. At pH 6.9, N_2 was the dominant product of denitrification, even after the first day of incubation. They also reported production of NO in soils with pH 4.6 and 5.4 that was consistent with the findings of Wijler and Dewiche (1954), who reported that considerable amounts of NO were evolved as a result of denitrification in soils at pH 4.9.

However, as suggested by Koskinen and Keeney (1982), it is difficult to evaluate the absolute effect of pH on the relative production of gaseous products of denitrification because of involvement of $[NO_2]^-$ and soil pH interactions that may affect production of N_2O and N_2. It is recognized that in more acid soils, chemical reactions of $[NO_2]^-$–N can also contribute to gaseous N production (Bollag *et al.*, 1973; Nelson, 1982).

7. Other Factors

In addition to the above-discussed environmental, substrate, and soil factors that affect N_2O emissions from soils, there are other factors that affect N_2O production which cannot be ascribed easily to any individual effect but are best explained by considering the total ecosystem. A combination of several environmental, soil, and agricultural practices, and perhaps plant factors, interact to influence such biological processes as mineralization, nitrification, and denitrification, which affect N_2O emissions.

For example, research by Aulakh *et al.* (1982) in Canadian prairies showed that gaseous N loss as (N_2O + N_2)-N was two to five times higher from summer-fallowed plots than from cropped plots. In a 2-yr study of various management practices on gaseous loss of N, Aulakh *et al.* (1983b) found that incorporation of legume residue followed by partial fallow decreased the gaseous loss of N compared to summer fallow. It was also found that soil moisture, mean air temperature, and inorganic N concentrations collectively accounted for 37 to 66% of the variations in

N_2O fluxes. The contribution of lower soil horizons to gaseous N loss was negligible.

Regarding the possible effect of ecosystems on N_2O fluxes, Goodroad and Keeney (1984b) observed major differences in N_2O emissions from forest soils. For example, it was found in a 2-yr study that significantly more N_2O was released from coniferous forest floors than from deciduous forest floors (Table 14). Differences in soil pH and N cycling processes might have contributed to these differences in N_2O fluxes. Usually low-N grassland ecosystems have been reported to emit small amounts of N_2O as compared to managed ecosystems (Mosier *et al.*, 1983; Goodroad and Keeney, 1984b).

Other factors, such as preincubation conditions (anaerobic and aerobic) and time of preincubation, also greatly affect N_2O evolution (Tiedje *et al.*, 1979; Letey *et al.*, 1980a, b). Letey *et al.* (1980a, b) found that little N_2O was evolved when the soil had been preincubated for 7 days under saturated conditions. There were relatively higher rates of N_2O evolution during the initial stages of incubation when the soil had been preincubated under air-dry or moist conditions. Nitrous oxide evolution decreased to nil with time. The ratio of N_2O to N_2 evolved was high initially and decreased rapidly with time mainly because of reduction of N_2O to N_2.

E. Other Processes Contributing to N_2O Production

In addition to biological reactions, chemodenitrification, in which volatile loss of N results from chemical reactions of $[NO_2]^-$ with organic reducing compounds, may contribute to production of N_2 and N_2O (Bremner and Blackmer, 1980; Nelson, 1982; Mosier *et al.*, 1983; Pang and Cho, 1984). Mosier *et al.* (1983) outlined five N_2O-producing processes and stated that nitrification and denitrification were the most important. However, in addition to chemodenitrification they listed dissimilatory $[NO_3]^-$ reduction to $[NH_4]^+$ and assimilatory $[NO_3]^-$ reduction (where N is incorporated into cell biomass) as processes that might contribute to N_2O evolution. Chemodenitrification may be of importance at times in soils with neutral to alkaline reaction where $[NO_2]$-$^-$ accumulation is a problem. However, research on N_2O emission from soils clearly establishes the importance of denitrification and nitrification as the main processes for N_2O production.

III. Mechanisms of N_2O Production in Soils

A. Chemical Reactions

Alternatively, the unidentified compound produced by oxidation of $[NH_4]^+$ or reduction of $[NO_2]^-$ may chemically react with organic

Table 14. Nitrous oxide emissions from University of Wisconsin Arboretum sites[a]

Site	Mean ($ng\ N_2O\text{-}N\ m^{-2}\ S^{-1}$)	Spatial S.D.[b]	Temporal S.D.[b]	Sampling dates	Total emissions ($kg\ N_2O\text{-}N\ ha^{-1}$)
1979					
Forests					
Deciduous	1.5	1.7	3.8	11 June–5 Nov.	0.19
Coniferous	7.7	9.5	18.1	11 June–5 Nov.	0.99
Prairie					
Burned	0.6	1.6	1.9	11 June–5 Nov.	0.09
Unburned	0.7	1.0	2.3	11 June–5 Nov.	0.10
Organic soils					
Drained marsh	18.0	13.8	45.8	11 June–5 Nov.	2.3
Undrained marsh	0.2	0.6	0.6	23 July–15 Nov.	0.02
Wet meadow	8.5	7.7	14.8	11 June–5 Nov.	1.1
1980					
Forests					
Deciduous	4.3	2.6	5.9	28 March–10 Nov.	0.80
Coniferous	10.1	4.9	8.5	28 March–10 Nov.	2.1
Prairie					
Burned	0.6	0.2	0.5	9 April–10 Nov.	0.11
Unburned	0.6	0.2	0.5	9 April–10 Nov.	0.10
Organic soils					
Drained marsh	41.3	35.4	134.7	28 March–10 Nov.	7.4
Undrained marsh	0.3	0.3	0.7	9 April–10 Nov.	0.06
Wet meadow	8.5	11.1	22.3	24 April–10 Nov.	1.4

[a]Reproduced from Goodroad and Keeney *J. Environ. Qual.* 13, 1984, 448–452, by permission of the Am. Soc. Agron., Crop Sci. Soc. Am., Soil Sci. Soc. Am.

[b]Standard deviations (S.D.) are pooled estimates between sample chambers (spatial) or between sampling dates (temporal).

materials to form N_2O as does $[NO_2]^-$. Accumulation of $[NO_2]^-$ is reported to occur in both acid and alkaline soils following application of alkaline-producing fertilizers, especially at high pH or at microsites where the fertilizer band is situated. Numerous factors, including low temperature, low soil buffering capacity, type and rate of fertilizer N, rate of nitrification, and soil pH, affect $[NO_2O]^-$ accumulation in soils (Chalk and Smith, 1983). Arnold (1954) and Wijler and Delwiche (1954) suggested that a Van Slyke-type chemical reaction may be responsible for production of N_2O through reaction of $[NO_2]^-$ with NH_2OH:

$$NH_2OH + NHO_2 \rightleftharpoons N_20 + 2\ H_20$$

Several subsequent reports also suggest that N_2O and N_2 are produced during chemical decomposition of hydroxylamine in soils (Bremner and Blackmer, 1980; Nelson, 1982; Chalk and Smith, 1983). Similarly, production of N gases, such as N_2 and N_2O, has been reported following the reaction of $[NO_2]^-$ with soil organic components. Discussions of the mechanism of N_2O formation via chemodenitrification are given in the reviews by Nelson (1982), Chalk and Smith (1983), and Hauck (1983).

B. Biological Reactions

Bremner and Blackmer (1980, 1981) have discussed the various factors affecting N_2O production in soils. It becomes evident from the cases they present that *Nitrosomonas* species have the capacity to produce nitrite reductase that reduces $[NO_2]^-$ to N_2O under anaerobic conditions or aerobically during concurrent oxidation of $[NH_4]^+$ to $[NO_2]^-$.

Work by Ritchie and Nicholas (1972, 1974) established that the conversion of $[NO_2]^-$ to N_2O (Yoshida and Alexander, 1970, 1971) by cells of *Nitrosomonas europaea* involves the production of an intermediate compound, nitroxyl, that dimerizes to form hyponitrous acid. Nitrous oxide production results from the decomposition of hyponitrous acid.

According to Bremner and Blackmer (1981), the mechanism of N_2O production during nitrification of $[NH_4]^+$ by *Nitrosomonas* spp. involves production of $[NO_2]^-$-N via NH_2OH and probably an unidentified compound. Nitrite is acted upon by nitrite reductase, resulting in evolution of N_2O.

Evidence presented by Payne (1973, 1981a, b), Ingraham (1981), Bryan (1981), Firestone (1982), and McKenney *et al.* (1982) suggests that the pathways of N_2O production employ the following route:

$$[NO_3]^- \xrightarrow[\text{reductase}]{\text{nitrate}} [NO_2[^- \xrightarrow[\text{reductase}]{\text{nitrite}} NO \xrightarrow[\text{reductase}]{\text{nitric oxide}} N_2O \xrightarrow[\text{reductase}]{\text{nitrous oxide}} N_2$$

However, there is still uncertainty about NO being an intermediate (Bryan, 1981; McKenney *et al.*, 1982) in the overall denitrification mechanism. The processes and mechanisms leading to N_2O and N_2 production are clearly established. The well-accepted pathway for N_2O and N_2 production is:

$$[NO_3]^- \rightarrow [NO_2]^- \rightarrow N_2O \rightarrow N_2$$

M. S. Smith and Tiedje (1979) suggested that there were two distinct phases of denitrification in soil. In phase I, the initial rate of denitrification was attributed to the activity of the preexisting denitrifying enzyme systems. This phase was little affected by C amendments and lasted for 1 to 3 hrs. Nitrous oxide concentrations reached measurable levels after 15 min and remained in easily measurable amounts during the rest of the phase I time. Chloramphenicol, an inhibitor of protein enzyme synthesis, did not decrease the rate of denitrification (N_2 and N_2O production) during phase I. Phase II of denitrification in soils was attained after 4 to 8 h of anaerobic incubation, and chloramphenicol inhibited the rate of denitrification in this phase. The rate of denitrification was linear because of a lack of significant growth of denitrifiers. Whereas phase I denitrification was found to be affected by the aeration state of the sample, phase II was not affected. It was suggested that although the denitrifying enzymes were present in relatively dry soils, they were not repressed until O_2 inhibition was eliminated. Evidence presented indicated that most of the N is lost through denitrification during the brief period of a few hours following an irrigation or a rainfall event. Further, the phase I denitrification rate was more likely to be related directly to field denitrification rates.

IV. Control of N_2O Emission from Soils

As discussed earlier, a range of soils and environmental factors affect N_2O emissions from soils via nitrification and denitrification processes. In addition to these factors, the production of N_2O is also greatly affected by soil, crop, and fertilizer N management practices (Mosier *et al.*, 1983). Studies have also shown that the cropping systems and the fertilization practices affect N_2O emissions from soils (Rolston, 1981; Mosier *et al.*, 1983). Some of the factors affecting N_2O production relating to soil ecosystems and management and native soil conditions are briefly covered here.

A. Effects of Management and Native Conditions on N_2O Emissions from Different Soil Ecosystems

Mosier *et al.* (1982) presented evidence to show that the source and rate of fertilizer N affect N_2O production in soils. In a field study of N_2O emissions from soil fertilized with 0, 56, 112, and 224 kg N ha^{-1} as

NH_4NO_3, it was found that N_2O emissions increased with the rate of N application. However, the increases in N_2O emissions with N rates were small and N_2O-N accounted for only about 0.4 to 1.5% of the fertilizer added (Mosier *et al.*, 1982). Similar results have been reported by many other researchers, indicating that the overall contribution of fertilizer N to N_2O production is rather small (Hutchinson and Mosier, 1979; Denmead *et al.*, 1979; McKenney *et al.*, 1978, 1980; Rolston *et al.*, 1980; Mosier and Hutchinson, 1981; Rolston, 1981; Cochran *et al.*, 1981; Seiler and Conrad, 1981; C. J. Smith *et al.*, 1981; Rice and Smith, 1982; Aulakh *et al*, 1982, 1983a, b; Armstrong, 1983) with the possible exception of organic soils (Terry *et al.*, 1981; Duxbury *et al.*, 1982; Mosier *et al.*, 1983) and vegetable crops heavily fertilized under irrigated conditions (Rolston, 1981; Rolston *et al.*, 1982).

Similarly, the source of fertilizer N also influences the fluxes of N_2O emissions from soils. For example, it has been observed that the N_2O emissions are larger from soils fertilized with anhydrous NH_3 than from those fertilized with $[NO_3]^-$ or $[NH_4]^+$ sources (Breitenbeck *et al.*, 1980; Bremner *et al.*, 1981; Cochran *et al.*, 1981; Mosier *et al.*, 1983).

Mosier *et al.* (1983) formulated a linear regression model of N_2O flux from soils describing the effect of fertilizer N rate on N_2O emissions, which can be represented by the following equation:

$$FN_2O = 3.88 + 0.024\,Ft$$

Where FN_2O is the N_2O flux expressed as g N ha^{-1} day^{-1} and Ft is the applied fertilizer expressed as kg N ha^{-1}.

The emissions of N_2O from soils have been characterized by a large efflux of N_2O for a short time following fertilizer N application, followed by declining rates approaching baseline levels, irrespective of the amounts of N applied. The period characterized by a large efflux of N_2O is reported to vary from 5 to 6 weeks when most of the N_2O emissions occur (Breitenbeck *et al.*, 1980; McKenney *et al.*, 1980; Bremner *et al.*, 1981; Mosier *et al.*, 1983).

Research summarized by Mosier *et al.* (1983) indicates that the amounts of N_2O evolved from the native grasslands were similar to amounts emitted from unfertilized cropped soils. The total amounts of N_2O-N emitted from these ecosystems annually amounted to about 1 kg N ha^{-1} and its source was apparently atmospheric deposition of N, which was about 10 kg N ha^{-1} annually.

Recent studies in Canada have shown that such management practices as summer fallowing, green manuring, and tillage affect the gaseous loss of N_2 and N_2O (Aulakh *et al.*, 1982, 1983a, b, 1984a).

1. Agricultural Soils

Recent field studies of N_2O emissions from agricultureal soils fertilized with chemical fertilizers or organic manures have been summarized by

Rolston (1981), Mosier *et al.* (1983), Ryden and Rolston (1983), and Goodroad *et al.* (1984). It becomes evident that N_2O emissions from arable soils are greatly influenced by fertilizer management practices. The form and rate of fertilizer applications influence N_2O emissions (Hutchinson and Mosier, 1979; McKenney *et al.*, 1978, 1980; Breitenbeck *et al.*, 1980; Bremner *et al.*, 1981; Cochran *et al.*, 1981; Seiler and Conrad, 1981; Mosier and Hutchinson, 1981; Mosier *et al.*, 1982, 1983). Also tillage and crop cover management (Aulakh *et al.*, 1982, 1983a, b) and irrigation and rainfall events that affect the water content of soils also affect N_2O emissions (Burford *et al.*, 1981; Rolston, 1981; Ryden, 1983; Ryden and Rolston, 1983; Goodroad *et al.*, 1984).

It has generally been found that the total N_2O emissions from agricultural soils fertilized with N are greatly influenced by the moisture regime. In dry regions with low rainfall activity, the annual emissions have been reported to be low and usually range from about 1 to 16 kg N_2O-N ha^{-1} (McKeeney *et al.*, 1978, 1980; Hutchinson and Mosier, 1979; Burford *et al.*, 1981; Aulakh *et al.*, 1982, 1983b). Nitrous oxide emission from N-fertilized and irrigated soils are reported to be of higher mangitude and can be as high as 20 to 42 kg N ha^{-1} (Ryden and Lund, 1980). Also, N_2O emissions from cropped or uncropped cultivated organic soils are larger than those from mineral soils (Terry *et al.*, 1981; Duxbury *et al.*, 1982).

Goodroad *et al.* (1984) measured the N_2O emissions from Wisconsin agricultural soils for two consecutive years and found that the total emissions were highest in the manure-amended and reduced tillage treatment (Table 15). Highest rate of N_2O emissions were often observed at the time of soil thaw. Soil pH (4.7, 5.1 and 6.7) had little effect on the emissions. The total emissions (in kg N_2O-N ha^{-1}) in 1980 under different treatments were: reduced tillage corn, 3.5 to 6.3; sludge, 1.6; manure, 6.1; alfalfa (*Medicago sativa*), 3.2; rye (*Secale cereale* L.), 1.6; straw, 2.2. These values are in the range of those reported in the literature from other arable agricultural soils (Mosier *et al.*, 1983).

Evidence presented by Mosier *et al.* (1983) suggests that the N_2O fluxes from the N-fertilized and irrigated fields in Colorado are larger than those from similar unfertilized sites under the same irrigation regimes. However, the N_2O-N emissions accounted for only 0.6 to 1.5% of the total N added.

With other agricultural soils, emissions of N_2O were highest from the vegetable cropland (Ryden and Lund, 1980) and drained organic soils used for cropping or left fallow (Duxbury *et al.*, 1982). However, their contribution to the overall global fluxes of N_2O may be small because of the small area covered by these soils.

Studies of N_2O emissions from agricultural arable soils so far reported, including both short-term and long-term measurements, suggest that generally less than 2.0% of the applied fertilizer N is lost as N_2O-N.

Table 15. Nitrous oxide emissions from agricultural soils in Wisconsin[a]

Site	(ng N_2O-N m^{-2} s^{-1})			Sampling dates	Total emissions[c] (kg N_2O-N ha^{-1})
	Mean	Spatial S.D.[b]	Temporal S.D.[b]		
1979					
Reduced tillage corn					
In the row	4.4	5.8	7.8	13 June–7 Nov.	0.6
Between the rows	2.4	2.8	5.2	13 June–7 Nov.	0.3
Organic amendments					
Sludge	1.9	1.0	2.2	21 June–20 Nov.	0.2
Manure	8.6	10.7	19.8	13 June–20 Nov.	1.2
Lime					
pH 4.7	2.8	1.4	3.8	13 June–20 Nov.	0.4
pH 5.1	1.7	0.9	2.7	13 June–20 Nov.	0.2
pH 6.7	1.6	1.4	2.7	13 June–20 Nov.	0.2
1980					
Reduced tillage corn					
In the row	21.5	23.7	30.8	22 April–29 Oct.	3.5
Between the rows	38.7	59.3	142.7	22 April–13 Nov.	1.6
Organic amendments					
Sludge	9.3	7.3	20.1	22 April–13 Nov.	1.6
Manure	27.7	48.1	79.1	2 April–15 Dec.	6.1
Lime					
pH 4.7	8.4	10.1	15.7	2 April–25 Oct.	1.5
pH 5.1	5.2	3.7	9.2	2 April–29 Oct.	0.9
pH 6.7	5.6	6.0	8.2	2 April–29 Oct.	1.0

[a]Adapted from Goodroad *et al. J. Environ. Qual.* 13, 1984, 557–561, by permission of the Am. Soc. Agron., Crop Sci. Soc. Am., Soil Sci. Soc. Am.
[b]Standard deviations (S.D.) are pooled estimates between sample chambers (spatial) and between sampling dates (temporal).
[c]Values obtained by summing the products of the daily emissions and time period in days over actual dates sampled.

Nitrous oxide emission rates are usually characterized by large temporal and spatial variability because of complex interactions between physical and biochemical factors that affect production and emission of N_2O from soils.

2. Forest Soils

There is lack of data on N_2O flux from forest soils ecosystems. A recent study by Goodroad and Keeney (1984b) suggested that there were large differences in N_2O emissions among various forest covers and that N_2O flux from replanted forest was on the same order of magnitude as that in agricultural systems (Table 14). It was also found that significantly more N_2O was emitted from coniferous than from deciduous forest floors. Perhaps N added through precipitation could be the source for these emissions. The average daily flux of N_2O-N from deciduous and coniferous forest floors were found to be 1.2 and 5.8 g ha^{-1} day^{-1}, respectively, and were 0.180 and 0.860 kg N_2O-N for the year. A recent study by Melillo *et al.* (1983) showed that N_2O emissions were highest from the forest floor of a 2-yr-old hardwood stand. Although the pH of the forest floors ranged from 3.5 to 3.9, there was considerable apparent denitrification loss of N as N_2O. The denitrification potential as measured by the acetylene blockage technique was highly correlated with the $[NO_3]^-$$-N$ content. The results suggested that recently cut forest systems have the potential to emit N_2O in amounts similar to those reported from agricultural soils.

Because forests occupy a significant area of land, their overall contribution to the global N_2O-N flux may be important, and there is need to obtain more data on N_2O emissions from forest soil ecosystems.

3. Prairies

Schultz *et al.* (1970) and Burford and Hall (1977) reported that the N_2O-N flux from unfertilized grasslands was small and ranged from 0.1 to 0.8 mg N m^{-2} day^{-1}. Burford and Hall (1977) found that fertilization of the grass sward increased the daily flux (range of 0.2 to 2 mg N_2O-N m^{-2}). Rolston *et al.* (1978) reported that N_2O emission rates from grass swards were the highest when they were heavily fertilized with organic manures and the soil water content was high. The N_2O flux in their study from grass sward ranged from 40 to 800 mg N m^{-2} day^{-1}. Ryden (1981), on the other hand, indicated that the daily N_2O flux from a British grassland soil fertilized with inorganic fertilizer N ranged from nil to 21 mg N m^{-2}. Soils very low in $[N0_3]^-$-N were found to act as sinks for atmospheric N_2O.

Mosier *et al.* (1981) studied N_2O emissions from a native short grass prairie and found that the daily flux averaged 2.3 g N ha^{-1} day^{-1} during summer. They found that fertilization of soils with urea to simulate grazing animal urine deposition increased the N_2O emission rates by a factor of three over the untreated soils. However, N_2O-N accounted for only 0.6% of the added N. They concluded that although N_2O emissions from the native grasslands were small, they were significantly larger than results obtained from uncultivated soils. Their estimate indicated that about 10% of total N input in the grasslands from biological N_2 fixation and atmospheric depositions was lost as N_2O. Sherlock and Goh (1983) noted that urine stimulated N_2O production in a grazed pasture relative to urea or $[NH_4]^+$. Large emissions of N_2O-N (up to 0.6 g ha^{-1} day^{-1}) occurred from sheep urine patches.

Denmead *et al.* (1979) measured N_2O emission from an unfertilized grass sward over 5 months and found that the emissions were largest in spring and increased with increase in the soil water and $[NO_3]^-$ contents. Nitrous oxide emission rates showed a diurnal cycle in response to soil surface temperatures. Peak emission rates occurred in the afternoon, with a minimum near sunrise. Average daily N_2O emissions from soils with soil water less than field capacity ranged from 0.6 to 2.5 mg N m^{-2}.

Christensen (1983) found that the N_2O flux from untreated soils under permanent grass during 5 months averaged 0.58 mg N m^{-2} day^{-1}. Application of cow dung slurry caused a surge in N_2O flux that increased up to 40-fold compared to untreated grassland. On the other hand, applications of NH_4NO_3 caused up to a five-fold increase in the N_2O flux during the same period (Table 16). These results, along with those reported by Rolston *et al.* (1978), suggest that grassland heavily fertilized with organic N has a greatly increased rate of N_2O emission.

Christensen (1983) reported that the N_2O flux often showed marked diurnal fluctuations, which were not solely associated with change in temperature but were also likely related to grass root activity and photosynthetic rates. A study by Blackmer *et al.* (1982) also showed that diurnal variability in rate of emissions of N_2O from Iowa soils was often not related to diurnal variability in soil temperature but was related to the solubility of N_2O in water induced by temperature change. The N_2O flux from the unfertilized grassland ranged from 0.5 to 1.4 mg N m^{-2} day^{-1} and increased up to 50 mg N m^{-2} day^{-1} when they were fertilized with organic N (Christensen, 1983).

Goodroad and Keeney (1984b) measured N_2O flux from Wisconsin prairies for 2 yr and found that the total N_2O-N emissions ranged from 0.09 to 0.11 kg N ha^{-1}. Nitrous oxide emissions were similar from burned and unburned prairies (Table 14). Seasonal patterns of N_2O emissions were markedly influenced by changes in soil temperature and soil water content. The average daily N_2O flux from uncultivated tall grass prairies amounted to 0.5 g N_2O-N ha^{-1} and was not significantly affected by removal of plant residues by burning.

Table 16. Cow slurry and ammonium nitrate-induced N_2O flux from soil under permanent grass[a,b]

Fertilizer/Slurry			
Treatment	Rate ($g\ N\ m^{-2}$)	Sampling date	N_2O-N flux ($mg\ N\ m^{-2}$)
None	—	1 May–10 July	43
	—	10 July–6 August	24
NH_4NO_3	10.0	1 May–10 July	132
	10.0	10 July–6 August	106
Slurry[c]	10.3/11.1	1 May–10 July	165
	15.9/11.9	10 July–6 August	770

[a]Reprinted with permission from *Soil Biol. Biochem.*, 15, S. Christensen, pp. 531-536; copyright 1983 Pergamon Press Ltd.

[b]Cow slurry and NH_4NO_3 were added to the plots on 30 April and 10 July 1981.

[c]Slurry was added on inorganic N/organic N basis.

4. Wetlands

Denmead *et al.* (1979) measured N_2O flux from a flooded rice field in New South Wales. The soils had about 40 kg NO_3-N ha^{-1} in the surface soils at the start of the experiment. Measurements were made continuously for 18 days after the field was flooded. Nitrous oxide fluxes showed a diurnal cycle in response to temperature of the paddy flood water. Between the second and 19th day of flooding, about 27 kg $[NO_3]^-$-N ha^{-1} was lost from the flood water. However, N_2O-N evolved was only 0.38 kg ha^{-1} and represented only about 1.4% of the apparent $[NO_3]^-$ loss. In another small-bay experiment in which $NaNO_3$ was added, N_2O production accounted for only 0.8% of the $[NO_3]^-$−N lost.

Similarly, a study of N_2O flux from a flooded rice paddy in the Philippines fertilized with $(NH_4)_2SO_4$ (80 kg N ha^{-1}) showed that N_2O emissions amounted to only 0.1% of the N applied (Freney *et al.*, 1981). Smith, C. J. *et al.* (1982) measured the rate of N_2O emissions from a flooded rice field in Louisiana fertilized with urea (0, 90, and 180 ga N ha^{-1}). Urea application increased the flux of N_2O-N, but the emissions were low and ranged from 90 to 197 g N ha^{-1} over 105 days. Nitrous oxide fluxes were found to be correlated with the content of exchangeable $[NH_4]^+$−N in soil and $([NO_3]^- + [NO_2]^-)$−N in the flood water. The average N_2O-N emissions from the unfertilized soil totaled 74 g N ha^{-1}, indicating that the fertilizer-induced emissions of N_2O represented only a very small proportion of the urea N applied (0.01 to 0.05%).

This brief discussion of the available information regarding N_20 emissions from flooded rice paddies clearly suggests that the N_20 emissions from these systems are very low and that fertilizer-induced N_20 emissions are also small, usually accounting for less than 1.4% of the fertilizer N.

B. Ecological Factors Affecting N_20 Emissions

It is evident that a large number of soil and environmental factors affect N_2O emission from soils. These include a number of physical and biochemical processes that directly or indirectly affect N_2O production or emission from soil.

In addition to these factors, the ecosystem as a whole likely has a profound effect on N_2O emissions. For example, recent studies have shown that there is a marked difference in N_2O emissions rate from different forest soil ecosystems representing different plant communities (Goodroad and Keeney, 1984b). There may be at least three factors that are affecting these emissions: (1) soil pH may be an important factor in controlling production of substrate for N_2O production via denitrification and nitrification. Perhaps immobilization of mineral N may be another factor controlling net availability of meneral N for nitrification and subsequent denitrification. (2) Deficiency of nutrients, such as P, may be implicated in N mineralization and nitrification, which affect N_2O emissions. (3) The possibility of some toxic substances being produced in a soil ecosystem, which then affects, nitrification and N mineralization, also needs investigation, because these "inhibitory substances" may be important in controlling the population of nitrifiers and denitrifiers involved in N_2O production.

All these ecological factors that may be affecting N_2O production in soils are speculative, for lack of supporting data, but may provide a good starting point for research to identify factors that affect N_2O emissions in different ecosystems.

C. Use of Chemicals to Control N_20 Emissions

Recent laboratory (Bremner and Blackmer, 1978; C. J. Smith and Chalk, 1980; Blackmer *et al.*, 1980; Aulakh *et al.*, 1984b) and field (Bremner *et al.* 1981; Magalhaes *et al.*, 1984) studies have shown that nitrapyrin reduces the rate of N_2O emission associated with the nitrification in soils. Field results showing the effect of nitrapyrin on anhydrous ammonia-induced N_2O emissions are given in Table 17, which clearly shows that nitrapyrin is effective in reducing N_2O emissions from both fall and spring applications of anhydrous ammonia. However, nitrapyrin has been

Table 17. Effects of nitrapyrin on N_2O emission from soil fertilized with anhydrous ammonia in fall or spring[a]

Treatment[b]	Amount of N_2O-N evolved (kg N ha^{-1}) in 167 days	
	Fall application	Spring application
None	0.16	0.43
AA	1.55	1.37
AA + NP	1.04	0.55

[a]From Bremner *et al., Geophys Res. Lett.* 8, 353–356, 1981, copyright by the American Geophysical Union.

[b]AA, anhydrous ammonia (180 kg N ha^{-1}); NP, nitrapyrin (0.56 kg ha^{-1}). Treatments were performed on 9 October, 1979 for the fall application and on 15 April, 1980 for the spring application.

found to have little direct effect on N_2O production via denitrification of $[NO_3]^-$-N (Bremner and Blackmer, 1980, 1981). This opens up a new avenue for the use of nitrification inhibitors to control N_2O emissions from soils.

There is a need to develop chemicals that can control N_2O emissions associated not only with nitrification but also with denitrification. A better understanding of the enzyme system involved in N_2O production should aid in developing chemicals that can block the specific enzymatic system which is responsible for N_2O production.

V. Summary

We have reviewed current literature on the processes that contribute to N_2O emissions from soils. It would appear that denitrification is the dominant source of N_2O evolution from soils. However, it is now well recognized that nitrification can release significant amounts of N_2O in some situations. Chemodenitrification as a mechanism of N_2O production in soils is well established under laboratory conditions. However, it is not possible at the present time to speculate how much nitrification and chemodenitrification contribute to the overall N_2O emissions from soils under field conditions. It is rather difficult to delineate the contributions of denitrification and nitrification to N_2O evolution from soils in the field because of the occurrence of anaerobic microsites in apparently aerobic soils, particularly ones with considerable native organic matter or carbon added externally as plant residues or manures.

The mechanism of N_2O production via nitrification of $[NH_4]^+$−N to $[NO_2]^-$-N is not clearly understood. It is clear, however, that N_2O can be formed from $[NO_2]^-$ via denitrification under anaerobic conditions or during oxidation of $[NH_4]^+$ to $[NO_2]^-$ under aerobic conditions. It is also not clearly understood whether N_2O is formed by chemical decomposition of hydroxylamine during nitrification or as a result of reaction between hydroxylamine and $[NO_2]^-$. It is clear, however that $[NO_2]^-$ accumulates in soils in situations where biological oxidation of $[NO_2]^-$ by *Nitrobacter* is at low ebb. There is a need to understand the significance of chemodenitrification in the field when alkaline-hydrolyzing fertilizers are band applied.

Research on N_2O emissions from native and managed ecosystems suggests that significant amounts of emissions occur under wet conditions from agricultural soils, particularly those fertilized with organic manures or having a high content of organic matter. The daily flux of N_2O-N can vary from as low as <0.01 to as high as 8 kg N ha^{-1}. The high fluxes of N_2O-N are not common in agricultural soils, with the exceptions of cultivated organic soils. Limited studies on forest soils suggest that fluxes of N_2O from these systems may be of an order of magnitude similar to those obtained from agricultural soils. Nitrous oxide emissions are greatly affected by the kind of forest stand and differences in N cycling processes. The potential for larger emissions of N_2O exists in recently cut forests.

The fluxes of N_2O from grasslands are quite variable and are affected by fertilization and other management practices. They range from 0.1 to 800 mg N m^{-2} day^{-1}. Mosier *et al.* (1981) reported an average daily flux of 2.3 g N ha^{-1} day^{-1} from a native short grass prairie in summer. They estimate that about 10% of total N inputs in the grasslands from biological N_2 fixation and atmospheric deposition is lost as N_2O annually. Nitrous oxide fluxes reported from flooded rice paddies are relatively small compared to other ecosystems. The fertilizer-induced N_2O emissions are also usually small from flooded rice soils. It would appear that under the anaerobic conditions found in flooded soils, the gaseous product of denitrification is dominantly N_2, because N_2O is readily reduced to N_2 under these conditions. These soils may act as sinks rather than sources for N_2O.

The best way of reducing N_2O emissions from soils is by increasing the efficiency of N in crop production. Nitrification inhibitors are capable of reducing N_2O production via nitrification or by chemodenitrification by slowing $[NO_2]^-$ accumulation. Development of chemicals that can retard denitrification and associated N_2O production will interest agronomists, soil scientists, and environmentalists.

Finally, modeling efforts are now needed to take stock of N_2O emissions from soils used for production of food, feed, and fiber crops with the available data sets as well as to learn about the kind of data sets needed for the modeling exercise prerequisite to prediction of global N_2O emissions.

References

Armstrong, A.S.B. 1983. Nitrous oxide emissions from two sites in Southern England during winter 1981/1982. *J. Sci. Food Agric.* 34:803–807.

Arnold, P.W. 1954. Losses of nitrous oxide from soil. *J. Soil Sci.* 5:116–128.

Aulakh, M.S., D.A. Rennie, and E.A. Paul. 1982. Gaseous nitrogen losses from cropped and summer fallowed soils. *Can. J. Soil Sci.* 62:187–195.

Aulakh, M.S., D.A. Rennie, and E.A. Paul. 1983a. The effect of various clover management practices on gaseous N losses and mineral N accumulation. *Can. J. Soil Sci.* 63:593–605.

Aulakh, M.S., D.A. Rennie, and E.A. Paul. 1983b. Field studies on gaseous nitrogen losses from soils under continuous wheat versus a wheat-fallow rotation. *Plant Soil* 75:15–28.

Aulakh, M.S., D.A. Rennie, and E.A. Paul. 1984a. Gaseous nitrogen losses from soils under zero-till as compared with conventional-till management systems. *J. Environ. Qual.* 13:130–136.

Aulakh, M.S., D.A. Rennie, and E.A. Paul. 1984b. Acetylene and N-Serve effects upon N_2O emissions from NH_4^+ and NO_3^- treated soils under aerobic and anaerobic conditions. *Soil Biol. Biochem.* 16:351–356.

Baalsrud, K., and K.S. Baalsrud. 1954. Studies on *Thiobacillus denitrificans. Arch. Mikrobiol.* 20:34–62.

Bailey, L.D. 1976. Effects of temperature and root denitrification in a soil. *Can. J. Soil Sci.* 56:79–87.

Balderston, W.L., B. Sherr, and W.J. Payne. 1976. Blockage by acetylene of nitrous oxide reduction in *Pseudomonas perfectomarinus. Appl. Environ. Microbiol.* 31:504–508.

Barbaree, J.M., and W.J. Payne. 1967. Products of denitrification by a marine bacterium as revealed by gas chromatography. *Mar. Biol.* 1:136–139.

Beijerinck, M.W., and D.C.J. Minkman. 1910. Bildung und Verbrauch von Stickoxydul. Zentralbl. Bakterial. II. Abt. 25:30–63 (Original not seen, cited in Payne, 1981b).

Blackmer, A.M., and J. M. Bremner. 1976. Potential of soil as a sink for atmospheric nitrous oxide. *Geophys. Res. Lett.* 3:739–742.

Blackmer, A.M., and J.M. Bremner. 1977. Nitrification of fertilizer nitrogen in soils as a source of atmospheric N_2O. *Agron. Abst.* p. 146.

Blackmer, A.M., and J.M. Bremner. 1978. Inhibitory effect of nitrate on reduction of N_2O to N_2 by soil microorganisms. *Soil Biol. Biochem.* 10:187–191.

Blackmer, A.M., J.M. Bremner, and E.L. Schmidt. 1980. Production of nitrous oxide by ammonia-oxidizing chemoautotrophic microorganisms in soil. *Appl. Environ. Microbiol.* 40:1060–1066.

Blackmer, A.M., S.G. Robbins, and J.M. Bremner. 1982. Diurnal variability in rate of emission of nitrous oxide from soils. *Soil Sci. Soc. Am. J.* 46:937–942.

Bleakley, B.H., and J.M. Tiedje. 1982. Nitrous oxide production by organisms other than nitrifiers or denitrifiers. *Appl. Environ. Microbiol.* 44:1342–1348.

Bollag, J.-M., S. Drzymala, and L.T. Kardos. 1973. Biological versus chemical nitrate decomposition in soil. *Soil Sci.* 116:44–50.

Breitenbeck, G.A., A.M. Blackmer, and J.M. Bremner. 1980. Effects of different nitrogen fertilizers on emission of nitrous oxide from soil. *Geophys. Res. Lett.* 7:85–88.

Bremner, J.M. 1978. Effects of soil processes on the atmospheric concentration of nitrous oxide. In: D.R. Nielsen and J.B. McDonald (eds.), *Nitrogen in the environment*. Vol. 1. Academic Press, New York, pp. 477–491.

Bremner, J.M., and A.M. Blackmer. 1978. Nitrous oxide: emission from soils during nitrification of fertilizer nitrogen. *Sicence* 199:295–296.

Bremner, J.M., and A.M. Blackmer. 1979. Effects of acetylene and soils water content on emission of nitrous oxide from soils. *Nature* (London) 280:380–381.

Bremner, J.M., and A.M. Blackmer. 1980. Mechanisms of nitrous oxide production in soils. In: P.A. Trudinger, M.R. Walter, and R.J. Ralph (eds.), *Biochemistry of ancient and modern environments*. Australian Academy of Science, Canberra, Australia, pp. 279–291.

Bremner, J.M., and A.M. Blackmer. 1981. Terrestrial nitrification as a source of atmospheric nitrous oxide. In: C.C. Delwiche (ed.), *Denitrification, nitrification, and atmospheric nitrous oxide*. John Wiley, New York, pp. 151–170.

Bremner, J.M., and K. Shaw. 1958. Denitrification in soil. II. Factors affecting denitrification. *J. Agric. Sci.* 51:39–52.

Bremner, J.M., G.A. Breitenbeck, and A.M. Blackmer. 1980. Effect of anhydrous ammonia fertilization on emission of nitrous oxide from soil. J. Environ. Qual. 10:77–80.

Bremner, J.M., G.A. Breitenbeck, and A.M. Blackmer. 1981. Effect of nitrapyrin on emission of nitrous oxide from soil fertilized with anhydrous ammonia. *Geophys. Res. Lett.* 8:353–356.

Broadbent, F.E., and F.E. Clark. 1965. Denitrification. In: W.V. Bartholomew and F.E. Clark (eds.), *Soil nitrogen*. Agronomy 9:344–359. Am. Soc. of Agron., Madison, Wisconsin.

Bryan, B.A. 1981. Physiology and biochemistry of denitrification. In: C.C. Delwiche (ed.), *Denitrification, nitrification, and atmospheric nitrous oxide*. John Wiley, New York, pp. 67–84.

Burford, J.R., and J.M. Bremner. 1975. Relationships between the denitrification capacities of soils and total water-soluble and readily decomposable soils organic matter. *Soil Biol. Biochem.* 7:389–394.

Burford, J.R., and K.C. Hall. 1977. Fluxes of nitrous oxide from a rye-grass sward. In: *Annual Rept. of Letcombe Lab. for 1976*, Agric. Res. Council, Wantage, U.K., pp. 85–88.

Burford, J.R., and R.J. Millington. 1968. Nitrous oxide in the atmosphere of a red brown earth. *Trans. 9th Int. Congr. Soil Sci.* 2:505–511.

Burford, J.R., and R.C. Stefanson. 1973. Measurements of gaseous losses of nitrogen from soils. *Soil Biol. Biochem.* 5:133–141.

Burford, J.R., R.J. Dowdell, and R. Crees. 1981. Emission of nitrous oxide to the atmosphere from direct-drilled and phoughed clay soils. *J. Sci. Food Agric.* 32:219–223.

Burth, I., G. Benckiser, and J.C.G. Ottow. 1982. N_2O-Freisetzung aus Nitrit (Denitrifikation) durch ubiquitare Pilze unter aeroben Bedingungen. *Naturwissenschaften* 69:598–599.

Cady, F.B., and W.V. Bartholomew. 1960. Sequential products of anaerobic denitrification in Norfolk soil material. *Soil Sci. Soc. Am. Proc.* 24:477–482.

Cady, F.B., and W.V. Bartholomew. 1961. Influence of low PO_2 on denitrification processes and products. *Soil Sci. Soc. Am. Proc.* 25:362–365.

Cady, F.B., and W.V. Bartholomew. 1963. Investigations of nitric oxide reactions in soils. *Soil Sci. Soc. Am. Proc.* 27:546–549.

CAST (Council for Agric. Sci. and Technol.). 1976. "Effect of increased nitrogen fixation on stratospheric ozone." Report 53, Iowa State Univ., Ames, Iowa.

Cates, R.L., Jr. 1983. Field and laboratory studies of nitrous oxide production in soils. Unpublished Ph.D. Thesis, University of Wisconsin-Madison, Wis. 281 pp.

Chalk, P.M., and C.J. Smith. 1983. Chemodenitrification. In: J.R. Freney and J.R. Simpson (eds.), *Gaseous loss of nitrogen from plant-soil systems. Developments in plant and soil sciences*. Vol. 9. Martinus Nijhoff, The Hague, pp. 65–89.

Cho, C.M. 1982. Oxygen consumption and denitrification kinetics in soil. *Soil Sci. Soc. Am. J.* 46:756–762.

Cho, C.M., and J.G. Mills. 1979. Kinetic formulation of the denitrification processes in soil. *Can. J. Soil Sci.* 59:249–257.

Christensen, S. 1983. Nitrous oxide emission from a soils under permanent grass: seasonal and diurnal fluctuations as influenced by manuring and fertilization. *Soil Biol. Biochem.* 15:531–536.

Cochran, V.L., L.F. Elliott, and R.I. Papendick. 1981. Nitrous oxide emissions from a fallow field fertilized with anhydrous ammonia. *Soil Sci. Soc. Am. J.* 45:307–310.

Cooper, G.S., and R.L. Smith. 1963. Sequence of products formed during denitrification in some diverse western soils. *Soil Sci. Soc. Am. Proc.* 27:659–662.

Corbet, A.S. 1935. The formation of hyponitroous acid as an intermediate compound in the biochemical oxidation of ammonia to nitrous acid. II. Microbiological oxidation. *Biochem. J.* 29:1086–1096.

Crutzen, P.J. 1970. The influence of nitrogen oxides on the atmospheric ozone content. *Quart. J. R. Meterol. Soc.* 96:320–325.

Crutzen, P.J. 1971. Ozone production rates in an oxygen-hydrogen-nitrogen oxide atmosphere. *J. Geophys. Res.* 76:7311–7328.

Crutzen, P.J. 1974. Estimates of possible variations in total ozone due to natural causes and human activities. *Ambio* 3:201–210.

Crutzen, P.J. 1981. Atmospheric chemical processes of the oxides of nitrogen, including nitrous oxide. In: C.C. Delwiche (ed.), *Denitrification, nitrification, and atmospheric nitrous oxide*. John Wiley, New York, pp. 17–44.

Daniel, R.M., K.W. Steele, and A.W. Limmer. 1980. Denitrification by rhizobia. A possible factor contributing to nitrogen losses from soils. *N.Z. Agric. Sci.* 14:109–112.

Delwiche, C.C. 1959. Production and utilization of nitrous oxide by *Pseudomonas denitrificans. J. Bacteriol.* 77:55–59.

Delwiche, C.C. (ed.). 1981. *Denitrification, nitrification and atmospheric nitrous oxide*. John Wiley, New York. 286 p.

Delwiche, C.C., and B.A. Bryan. 1976. Denitrification. *Ann. Rev. Microbiol.* 30:241–262.

Denmead, O.T., J.R. Freney, and J.F. Simpson. 1979. Studies of nitrous oxide emission from a grass sward. *Soil Sci. Soc. Am. J.* 43:726–728.

Dowdell, R.J., and K.A. Smith. 1974. Field studies of the soil atmosphere. II. Occurrence of nitrous oxide. *J. Soil Sci.* 25:231–238.

Duxbury, J.M., D.R. Bouldin, R.E. Terry, and R.L. Tate, III. 1982. Emissions of nitrous oxide from soils. *Nature* (London) 298:462–464.

Elkins, J.W., S.C., Wofsy, M.B. McElroy, C.E. Kolb, and W.A. Kaplan. 1978. Aquatic sources and sinks for nitrous oxide. *Nature* (London) 275:602–604.

Findlay, W.I., and D.J. McKenney. 1979. Direct measurement of nitrous oxide flux from soil. *Can. J. Soil Sci.* 59:413–421.

Firestone, M.K. 1982. Biological denitrification. In: F.J. Stevenson *et al.* (eds.) Nitrogen in agricultural soils. *Agronomy* 22:289–236. Am. Soc. Agron., Madison, Wisconsin.

Firestone, M.K., and J.M. Tiedje. 1979. Temporal changes in nitrous oxide and denitrogen from denitrification following on set of anaerobiosis. *Appl. Environ. Microbiol.* 38:673–679.

Firestone, M.K., R.B. Firestone, and J.M. Tiedje, 1980. Nitrous oxide from soil denitrification: factors controlling its biological production. *Science* 208:749–751.

Focht, D.D. 1974. The effect of temperature, pH and aeration on the production of nitrous oxide and gaseous nitrogen—a zero order kinetic model. *Soil Sci.* 118:173–179.

Focht, D.D. 1978. Methods for analysis of denitrification in soils. In: D.R. Nielsen and J.B. McDonald (eds.), *Nitrogen in the environment.* Vol. 2. *Soil-plant relationships.* Academic Press, New York, pp. 433–490.

Focht, D.D., and W. Verstraete. 1977. Biochemical ecology of nitrification and denitrifictaion. In: M. Alexander (ed.), *Advances in microbial ecology.* Vol. 1. Plenum Press, New York, pp. 135–214.

Focht, D.D., L.H. Stolzy, and B.D. Meek. 1979. Sequential reduction of nitrate and nitrous oxide under field conditions as brought about by organic amendments and irrigation management. *Soil Biol. Biochem.* 11:37–46.

Freney, J.R., and J.R. Simpson (eds.). 1983. Gaseous loss of nitrogen from plant-soil systems. In: *Development in plant and soil sciences.* Vol. 9. Martinus Nijhoff, The Hague, p. 317.

Freney, J.R., O.T. Denmead, and J.R. Simpson. 1978. Soil as a source or sink for atmospheric nitrous oxide. *Nature* (London) 273:530–532.

Freney, J.R., O.T. Denmead, and J.R. Simpson. 1979. Nitrous oxide emission from soils at low moisture contents. *Soil Biol. Biochem.* 11:167–173.

Freney, J.R., O.T. Denmead, I. Watanabe, and E.T. Craswell. 1981. Ammonia and nitrous oxide losses following applications of ammonium sulfate to flooded rice. *Aust. J. Agric. Res.* 32:37–45.

Garcia, J.L. 1975a. La denitrification dans les sols. *Bull. Inst. Pasteur Paris* 73:167–193.

Garcia, J.L. 1975b. Effet rhizosphere du riz sur la denitrification. *Soil Biol. Biochem.* 7:139–141.

Goodroad, L.L., and D.R. Keeney. 1984a. Nitrous oxide production in aerobic soils under varying pH, temperature, and water content. *Soil Biol. Biochem.* 16:39–43.

Goodroad, L.L., and D.R. Keeney. 1984b. Nitrous oxide emission from forest, marsh, and prairie ecosystems. *J. Environ. Qual.* 13:448–452.

Goodroad, L.L., and D.R. Keeney, 1985. Site of nitrous oxide production in field soils. *Biol. Fert. Soils* 1:3–7.

Goodroad, L.L., D.R. Keeney, and L.A. Peterson. 1984. Nitrous oxide from agricultural soils in Wisconsin. *J. Environ. Qual.* 13:557–561.

Goreau, T.J., W.A. Kaplan, S.C. Wofsy, M.B. McElroy, F.W. Valois, and S.W. Watson. 1980. Production of NO_2^- and N_2O by nitrifying bacteria at reduced concentrations of oxygen. *Appl. Environ. Microbiol.* 40:526–532.

Hahn, J. 1981. Nitrous oxide in the oceans. In: C.C. Delwiche (ed.), *Denitrification, nitrification, and atmospheric nitrous oxide*. John Wiley, New York, pp. 191–240.

Hauck, R.D. 1983. Agronomic and technological approaches to minimizing gaseous nitrogen losses from croplands. In: J.R. Freney and J.R. Simpson (eds.), *Gaseous loss of nitrogen from plant-soil systems. Development in plant and soil sciences*, Vol. 9. Martinus Nijhoff, The Hague; pp. 285–312.

Hauck, R.D., and S.W. Melsted. 1956. Some aspects of the problem of evaluating denitrification in soils. *Soil Sci. Soc. Am. Proc.* 20:361–364.

Hollocher, T.C. 1982. The pathway of nitrogen and reductive enzymes of denitrification. *Antonie Van Leeuwenhoek* 48:531–544.

Hutchinson, G.L., and A.R. Mosier. 1979. Nitrous oxide emissions from an irrigated corn field. *Science* 205:1125–1127.

Hutchinson, G.L., and A.R. Mosier. 1981. Improved soil cover method for field measurement of nitrous oxide flux. *Soil Sci. Soc. Am. J.* 45:311–316.

Hynes, R.K., and R. Knowles. 1978. Inhibition by acetylene of ammonia oxidation in *Nitrosomonas europaea. FEMS Microbiol. Lett.* 4:319–321.

Ingraham, J.L. 1981. Microbiology and genetics of denitrifiers. In: C.C. Delwiche (ed.), *Denitrification, nitrification, and atmospheric nitrous oxide*. John Wiley, New York, pp. 45–65.

Ishaque, M., and M.I.H. Aleem. 1973. Intermediates of denitrification in the chemoautotroph *Thiobacillus denitrificans. Arch. Mikrobiol.* 94:269–282.

Johnson, H.S. 1971. Reduction of stratospheric ozone by nitrogen oxide catalysts from SST exhaust. *Science* 173:517–522.

Johnston, H.S. 1977. Analysis of independent variables in the perturbation of stratospheric ozone by nitrogen fertilizers. *J. Geophys. Res.* 82:1767–1772.

Keeney, D.R. 1982. Nitrogen management for maximum efficiency and minimum pollution. In: F.J. Stevenson *et al.* (ed.), *Nitrogen in agricultural soils. Agronomy* 22:605–649. Am. Soc. Agron., Madison, Wisconsin.

Keeney, D.R. 1986. Critique of the acetylene blockage technique for field measurement of denitrification. In: R.D. Hauck and R. W. Weaver (eds.) *Field measurement of dinitrogen fixation and denitrification*. Soil Sci. Soc. Am. Spec. Publ. 18: pp. 103–115. Soil Sci. Soc. Am., Madison, Wisconsin.

Keeney, D.R., I.R. Fillery, and G.P. Marx. 1979. Effect of temperature on the gaseous nitrogen products of denitrification in a silt loam. *Soil Sci. Soc. Am. J.* 43:1124–1128.

Keeney, D.R., K.L. Sahrawat, and S.S. Adams. 1985. Carbon dioxide concentration in soil: Effects on nitrification, denitrification and associated nitrous oxide production. *Soil Biol. Biochem.* 17:571–573.

Knowles, R. 1981. Denitrification. In: E.A. Paul and J.M. Ladd (eds.), *Soil Biochemistry*. Vol. 5. Marcel Dekker, New York, pp. 323–369.

Koskinen, W.C., and D.R. Keeney. 1982. Effects of pH on the rate of gaseous production of denitrification in a silt loam soil. *Soil Sci. Soc. Am. J.* 46:1165–1167.

Letey, J., N. Valoras, A. Hadas, and D.D. Focht. 1980a. Effect of air-filled porosity, nitrate concentration and time on the ratio of N_2O/N_2 evolution during denitrification. *J. Environ. Qual.* 9:227–321.

Letey, J., A. Hadas, M. Valoras, and D.D. Focht. 1980b. Effect of preincubation treatment on the ratio of N_2O/N_2 evolution. *J. Environ. Qual.* 9:323–235.

Letey, J., N. Valoras, D.D. Focht, and J.C. Ryden, 1981. Nitrous oxide production and reduction during denitrification as affected by redox potential. *Soil Sci. Soc. Am. J.* 45:727–730.

Liu, S.C., R.J. Cicerone, and T.M. Donahue. 1977. Sources and sinks of atmospheric N_2O and possible ozone reduction due to industrial fixed nitrogen fertilizers. *Tellus* 29:251–263.

Magalhaes, A.M.T., P.M. Chalk, and W.M. Strong. 1984. Effect of nitrapyrin on nitrous oxide emission from fallow soils fertilized with anhydrous ammonia. *Fert. Res.* 5:411–421.

Matsurbara, T., and T. Mori. 1968. Studies on denitrification. IX. Nitrous oxide, its production and reduction to nitrogen. *J. Biochem.* (Tokyo) 64:863–871.

McElroy, M.B., S.C. Wofsy, and Y.L. Yung. 1977. The nitrogen cycle: perturbations due to man and their impact on atmospheric N_2O and O_3. *Phil. Trans. R. Soc. London, B. Biol Sci.* 277:159–181.

McGarity, J.W. 1962. Effect of freezing of soil on denitrification. *Nature* (London) 196:1342–1343.

McGarity, J.W., and R.D. Hauck. 1969. An aerometric apparatus for the evaluation of gaseous nitrogen transformations in field soils. *Soil Sci.* 108:335–344.

McKenney, D.J., D.L. Wade, and W.I. Findlay. 1978. Rates of N_2O evolution from N-fertilized soil. *Geophys. Res. Lett.* 5:777–780.

McKenney, D.J., K.F. Shuttleworth, and W.I. Findlay, 1980. Nitrous oxide evolution rates from fertilized soils; effects of applied nitrogen. *Can. J. Soil Sci.* 60:429–438.

McKenney, D.J., K.F. Shuttleworth, J.R. Vriesacker, and W.I. Findlay. 1982. Production and loss of nitric oxide from denitrification in anaerobic Brookston clay. *Appl. Environ. Microbiol.* 43:534–541.

McKenzie, E., and L.T. Kurtz. 1976. Effect of pretreatment on loss of nitrogen-15-labeled fertilizer from waterlogged soil during incubation. *Soil Sci. Soc. Am. J.* 40:534–537.

Melillo, J.M., J.D. Aber, P.A. Steudler, and J.P. Schimel. 1983. Denitrification potential in a successional sequence of northern hardwood forest stands. *Environ. Biogeochem. Ecol. Bull.* (Stockholm) 35:217–228.

Minami, K., and S. Fukushi. 1983. Effects of phosphate and calcium carbonate application on emission of N_2O from soils under aerobic conditions. *Soil Sci. Plant Nutr.* 29:517–524.

Miyata, M. 1971. Studies on denitrification. XIV. The electron donating system in the reduction of nitric oxide and nitrate. *J. Biochem.* (Toyko) 70:205–213.

Mosier, A.R., and G.L. Hutchinson. 1981. Nitrous oxide emissions from cropped fields. *J. Environ. Qual.* 10:169–173.

Mosier, A.R., M. Stillwell, W.J. Parton, and R.G. Woodmansee. 1981. Nitrous oxide emissions from a native short grass prairie. *Soil Sci. Soc. Am. J.* 45:617–619.

Mosier, A.R., F.L. Hutchinson, B.R. Sabey, and J. Baxter. 1982. Nitrous oxide emissions from barley plots treated with ammonium nitrate or sewage sludge. *J. Environ. Qual.* 11:78–81.

Mosier, A.R., W.J. Parton, and G.L. Hutchinson. 1983. Modelling nitrous oxide evolution from cropped and native soils. *Environ. Biogeochem. Ecol. Bull.* (Stockholm) 35:229–241.

Myers, R.J.K., and J.W. McGarity. 1971. Factors influencing high denitrifying activity in the subsoil of solodized solonetz. *Plant Soil* 35:145–160.

Najjar, V.A., and M.B. Allen. 1954. Formation of nitrogen, nitrous oxide and nitric oxide by extracts of denitrifying bacteria. *J. Biol. Chem.* 206:209–214.

Nelson, D.W. 1982. Gaseous losses of nitrogen other than through denitrification. In: F.J. Stevenson *et al.* (eds.), Nitrogen in agricultural soils. Agronomy 22:327–363. Am. Soc. Agron., Madison, Wisconsin.

Nelson, D.W., and J.M. Bremner. 1970. Gaseous products of nitrite decomposition in soils. *Soil Biol. Biochem.* 2:203–215.

Nommik, H. 1956. Investigations on denitrification in soil. *Acta Agric. Scand.* 6:195–228.

O'Hara, G.W., R.M. Daniel, and K.W. Steele. 1983. Effect of oxygen on the synthesis, activity and breakdown of the *Rhizobium* denitrification system. *J. Gen. Microbiol.* 129:2405–2412.

O'Hara, G.W., R.M. Daniel, K.W. Steele, and P.M. Bonish. 1984. Nitrogen losses from soils caused by *Rhizobium*-dependent denitrification. *Soil Biol. Biochem.* 16:429–431.

Pang, P.C.K., and C.M. Cho. 1984. Oxygen consumption and denitrification activity of a conifer forest soil profile. *Soil Sci. Soc. Am. J.* 48:393–399.

Pastor, J., J.D. Aber, C.A. McClaugherty, and J.M. Melillo. 1984. Above-ground production and N and P cycling along a nitrogen mineralization gradient on Blackhawk Island, Wisconsin. *Ecology* 65:256–268.

Patten, D.K., J.M. Bremner, and A.M. Blackmer. 1980. Effects of drying and air-dry storage of soils on their capacity for denitrification nitrate. *Soil Sci. Soc. Am. J.* 44:67–70.

Payne, W.J. 1973. Reduction of nitrogenous oxide by microorganisms. *Bacteriol. Rev.* 37:409–452.

Payne, W.J. 1981a. The status of nitric oxide and nitrous as intermediates in denitrification. In: C.C. Delwiche (ed.), *Denitrification, nitrification, and atmospheric nitrous oxide.* John Wiley, New York, pp. 85–103.

Payne, W.J. 1981b. *Denitrification*. John Wiley, New York. 214 pp.

Pichinoty, F., J.L. Garicia, M. Mandel, C. Job, and M. Durand. 1978. Isolement de bacteries utilisant en anaerobiose l'oxyde nitrique cdomme accepteur d'electrons respirtoire. *C.R. Acad. Sci.*, Ser. D 286:1403–1405.

Pilot, L., and W.H. Patrick, Jr. 1972. Nitrate reduction in soils: effect of soil moisture tension. *Soil Sci.* 114:312–316.

Rice, C.W., and M.S. Smith. 1982. Denitrification in no-till and plowed soils. *Soil Sci. Soc. Am. J.* 46:1168–1173.

Ritchie, G.A.F., and D.J.D. Nicholas. 1972. Identification of the sources of nitrous oxide produced by oxidative and reductive processes in *Nitrosomonas europaea. Biochem. J.* 126:1181–1191.

Ritchie, G.A.F., and D.J.D. Nicholas. 1974. The partial characterization of purified reductase and hydroxylamine oxidase from *Nitrosomonas europaea Biochem. J.* 138:471–480.

Robertson, G.P., and J.M. Tiedje. 1984. Denitrification and nitrous oxide production in successional and old-growth Michigan forests. *Soil Sci. Soc. Am. J.* 48:383–389.

Rolston, D.E. 1981. Nitrous oxide and nitrogen gas production in fertilizer loss.

In: C.C. Delwiche (ed.), *Denitrification, nitrification, and atmospheric nitrous oxide.* John Wiley, New York, pp. 127–149.

Rolston, D.E., M. Fried, and D.A. Goldhamer. 1976. Denitrification measured directly from nitrogen and nitrous oxide gas fluxes. *Soil Sci. Soc. Am. J.* 40:259–266.

Rolston, D.E., D.L. Hollman, and D.W. Toy. 1978. Field measurement of denitrification. I. Flux of N_2 and N_2O. *Soil Sci. Soc. Am. J.* 42:863–869.

Rolston, D.E., A.N. Sharpley, D.W. Toy, D.L. Hoffman, and F.E. Broadbent. 1980. "Denitrification as affected by irrigation frequency of a field soil." EPA-600/2-80-066. U.S. Environmental Protection Agency, Ada, Oklahoma.

Rolston, D.W., A.N. Sharpley, D.W. Toy, and F.E. Broadbent. 1982. Field measurement of denitrification: III. Rates during irrigation cycles. *Soil Sci. Soc. Am. J.* 46:289–296.

Ryden, J.C. 1981. N_2O exchange between a grassland soil and the atmosphere. *Nature* (London) 292:235–237.

Ryden, J.C. 1983. Denitrification loss from a grassland soils in the field receiving different rates of nitrogen as ammonium nitrate. *J. Soil Sci.* 34:355–365.

Ryden, J.C., and L.J. Lund. 1980. Nature and extent of directly measured denitrification losses from some irrigated vegetable crop production units. *Soil Sci. Soc. Am. J.* 44:505–511.

Ryden, J.C., and D.E. Rolston. 1983. The measurement of denitrification. In: J.R. Freney and J.F. Simpson (eds.), *Gaseous loss of nitrogen from plant-soil systems. Developments in plant and soil sciences.* Vol. 9. Martinus Nijhoff, The Hague, pp. 91–132.

Ryden, J.C., L.J. Lund, and D.D. Focht. 1978. Direct-in-field measurement of nitrous oxide flux from soils. *Soil Sci. Soc. Am. J.* 42:731–737.

Ryden, J.C., L.J. Lund, and D.D. Focht. 1979. Direct measurement of denitrification loss from soils: development and application of field methods. *Soil Sci. Soc. Am. J.* 43:110–118.

Sahrawat, K.L., D.R. Keeney, and S.S. Adams. 1985. Rate of aerobic nitrogen transformations in six acid climax forest soils and the effect of phosphorus and $CaCO_3$. *Forest Sci.* 31:680–684.

Schroeder, E.D. 1981. Denitrification in wastewater management. In: C.C. Delwiche (ed.), *Denitrificaton, nitrification, and atmospheric nitrous oxide.* John Wiley, New York, pp. 105–125.

Schultz, K., C. Junge, R. Beck, and B. Albrecht. 1970. Studies of atmospheric N_2O. *J. Geophys. Res.* 75:2230–2246.

Seiler, W., and R. Conrad. 1981. Field measurement of natural and fertilizer-induced N_2O release rates from soils. *Air Polln. Contr. Assn. J.* 31:767–772.

Sherlock, R.R., and K.M. Goh. 1983. Initial emission of nitrous oxide from sheep urine applied to pasture soil. *Soil Biol. Biochem.* 15:615–617.

Smith, C.J., and P.M. Chalk. 1980. Gaseous nitrogen evolution during nitrification of ammonia fertilizer and nitrite transformations in soils. *Soil Sci. Soc. Am. J.* 44:277–282.

Smith, C.J., and W.H. Patrick, Jr. 1983. Nitrous oxide emission as affected by alternate anaerobic and aerobic conditions from soil suspensions enriched with ammonium sulfate. *Soil Biol. Biochem.* 15:693–697.

Smith, C.J., M. Brandon, and W.H. Patrick, Jr. 1982. Nitrous oxide emission following urea-N fertilization of wetland rice. *Soil Sci. Plant Nutr.* 28:161–171.

Smith, C.J., R.L. Chen, and W.H. Patrick, Jr. 1981. Nitrous oxide emissions from sinulated overland flow water treatment systems. *Soil Biol. Biochem.* 13:275–278.

Smith, C.J., M.F. Wright, and W.H. Patrick, Jr. 1983. The effect of soil redox potential and pH on the reduction and production of nitrous oxide. *J. Environ. Qual.* 12:186–188.

Smith, K.A. 1980. A model of the extent of anaerobic zones in aggregated soils, and its potential application to estimates of denitrification. *J. Soil Sci.* 31:263–277.

Smith, M.S., and J.M. Tiedje. 1979. Phases of denitrification following oxygen depletion in soil. *Soil Biol. Biochem.* 11:261–276.

Smith, M.S., and J.M. Tiedje. 1980. Growth and survival of antibiotic-resistant denitrifier strains in soil. *Can. J. Microbiol.* 26:854–856.

Smith, M.S., and K. Zimmerman. 1981. Nitrous oxide production by nondenitrifying soil nitrate reducers. *Soil Sci. Soc. Am. J.* 45:865–871.

Smith, M.S., M.K. Firestone, and J.M. Tiedje. 1978. The acetylene inhibition method for short-term measurement of soil denitrification and its evaluation using nitrogen-13. *Soil Sci. Soc. Am. J.* 42:611–616.

Soil Science Society of America, Terminology Committees. 1979. *Glossary of soil science terms.* Soil Sci. Soc. Am., Madison, Wisconsin, p. 5.

Stefanson, R.C. 1973. Evolution patterns of nitrous oxide and nitrogen in sealed soil-plant systems. *Soil Biol. Biochem.* 5:167–169.

Suzuki, S. 1912. Uber die enstehungder Stickoxyde in denitrifikations-prozess. I. Prufung, Bestimung und Voskomen des Stickoxyduls in den Garungsgasen. *Zentrabl. Bakteriol. Parasitenkd.* II Abt. 31:27–49.

Terry, R.E., and R.L. Tate, III. 1980. The effect of nitrate on nitrous oxide reduction in organic soils and sediments. *Soil Sci. Soc. Am. J.* 44:744–747.

Terry, R.E., R.L. Tate, III, and J.M. Duxbury. 1981. Nitrous oxide emissions from drained cultivated organic soils of South Florida. *Amer. Poll. Contr. Assn. J.* 31:1173–1176.

Tiedje, J.M. 1982. Denitrification. In: A.L. Page (ed.), *Methods of soil analysis*, Part 2. Agronomy 9:1011–1026. Am. Soc. Agron., Soil Sci. Soc. Am., Madison, Wisconsin.

Tiedje, J.M., R.B. Firestone, M.K. Firestone, M.F. Betlach, M.S. Smith, and W.H. Caskey. 1979. Methods for the production and use of nitrogen-13 in studies of denitrificaton. *Soil Sci. Soc. Am. J.* 43:709–716.

Trudinger, P.A., M.R. Walter, and R.J. Ralph. (eds.). 1980. Biochemistry of ancient and modern environments. Australian Academy of Science, Canberra, Australia.

Walter, H.M., D.R. Keeney, and I.R. Fillery. 1979. Inhibition of nitrification by acetylene. *Soil Sci. Soc. Am. J.* 43:195–196.

Webster, C.P., and R.J. Dowdell. 1982. Nitrous oxide emission from permanent grass swards. *J. Sci. Food Agric.* 33:227–230.

Wijler, J., and C.C. Delwiche. 1954. Investigations on the denitrification process in soil. *Plant Soil* 5:155–169.

Yoshida, T., and M. Alexander. 1970. Nitrous oxide formation by *Nitrosomonas europea* and heterotrophic microorganisms. *Soil Sci. Soc. Am. Proc.* 34:880–882.

Yoshida, T., and M. Alexander. 1971. Hydroxylanine oxidation by *Nitrosomonas europaea*. *Soil Sci.* 111:307–312.

Yoshinari, T., and R. Knowles. 1976. Acetylene inhibiton of nitrous oxide reduction by denitrifying bacteria. *Biochem. Biophys. Res. Commun.* 69:705–710.

Yoshinari, T., R. Hynes, and R. Knowles. 1977. Acetylene inhibition of nitrous oxide reduction and measurement of denitrification and nitrogen fixation in soil. *Soil Biol. Biochem.* 9:177–183.

Yung, Y.L., and W.C. Wang, and A.A. Lacis. 1976. Greenhouse effects due to nitrous oxide. *Geophys. Res. Lett.* 3:619–621.

Zablotowicz, R.M., D.L. Eskew, and D.D. Focht. 1978. Denitrification in *Rhizobium*. *Can. J. Microbiol*. 24:757–760.

Characterization and Modeling of Chemical Transfer to Runoff*

L. R. Ahuja*

I. Introduction

The release and migration of nutrients, pesticides, and other chemicals from agricultural lands is both an economic loss and a threat to the quality of our surface and ground waters. This problem was recognized more than two decades ago, and a large number of investigations on the subject have taken place since the early 1970s. In the United States, the research has been stimulated by the Federal Water Pollution Control Act Amendments of 1972 (PL 92-500), which specify that the Administrator of the U.S. Environmental Protection Agency shall, in cooperation with State and other Federal agencies, provide guidelines for identifying and evaluating the nature and extent of nonpoint sources of pollution.

---

*Contribution of the Water Quality and Watershed Research Laboratory, USDA-ARS, Durant, OK, USA

This chapter was prepared by the author as part of his official duties as a U.S. Government employee; copyright is not claimed.

Section 208 of that law specifies that agriculturally related sources be identified and that procedures and methods be developed to control such sources. A comprehensive evaluation of the problems and development of guidelines followed this law (Stewart *et al.*, 1975, 1976). Agricultural activities have been implicated as significant contributors to the eutrophication of several water bodies, ranging from small lakes, such as Cobbossee Lake in Maine (U.S. Environmental Protection Agency, 1980), to large ones, such as the Great Lakes (Drynan, 1980) and coastal estuarine waters (Duda, 1982). Pesticide contents of these waters are also of serious concern in many cases.

The majority of the experimental investigations conducted so far have been concerned with the determination of overall loss of nutrients and pesticides in runoff from different land uses, cropping systems, management practices, and climatic conditions (e.g., Menzel *et al.*, 1978; Johnson *et al.*, 1979; Monke *et al.*, 1981; McDowell *et al.*, 1981, 1984; Miller *et al.*, 1982; Hubbard *et al.*, 1982; Scheppers *et al.*, 1982; Duda and Finn, 1983; Wu *et al.*, 1983). The losses of agricultural chemicals in runoff are generally less than 5% of the amounts applied (Olness *et al.*, 1975; Wauchope, 1978). However, the losses can be much higher under certain soil and rainfall conditions. Schuman *et al.* (1973a, b) reported an average yearly loss of 15% for N from an Iowa corn (*Zea mays*) watershed. McDowell *et al.* (1980) found that the equilibrium phosphorus concentra-Mississippi Delta watershed. Pesticide losses as high as 10 to 15% of the amount applied have been measured where a heavy rainfall occurred soon after application (Baker *et al.*, 1978; Baker and Johnson, 1979; Baker and Laflen, 1979). Most of the loss in runoff generally occurs in a few heavy rainstorms, and the concentrations of chemicals in runoff can also be high during those storms (Edwards *et al.*, 1980; Hubbard *et al.*, 1982).

Effects of conservation tillage on chemical losses in runoff have been of special interest. It has been found that whereas the conservation tillage greatly reduces soil erosion and the loss of chemicals associated with sediment, as compared with conventional tillage, the loss of chemicals in solution form may be increased under certain conditions (Romkens *et al.*, 1973; Barisas *et al.*, 1978; Siemens and Oschwald, 1978; Baker *et al.*, 1978; Baker and Johnson, 1979; Johnson *et al.*, 1979; McDowell and McGregor, 1980; Laflen and Tabatabai, 1984). The major reason for this result is that the chemicals are minimally incorporated into the soil under conservation tillage. However, the leaching of nutrients from plant residues left on the soil surface under conservation tillage also contributes to their increased losses in runoff (McDowell and McGregor, 1984; Schreiber, 1985; Schreiber and McDowell, 1985). A related important finding is that for most of the common herbicides a large percentage (up to 90%) of the loss in overland runoff is in solution form (Baker *et al.*, 1978; Baker and Johnson, 1979; Baker *et al.*, 1982; Wu *et al.*, 1983). A substantial proportion of nitrogen (8–80%) and phosphorus (7–30%) losses in runoff may also be in solution form (Schuman *et al.*, 1973a, b; Kissel *et al.*, 1976; Menzel *et al.*, 1978; Hubbard *et al.*, 1982; McDowell *et al.*, 1984). Besides overland flow, the shallow subsurface flow of water can be an important contributor of soluble chemicals to runoff where the topsoil is underlain by an impeding layer (Owens *et al.*, 1983; Hubbard and Sheridan, 1983; Wu *et al.*, 1983). Because it is not practical to measure chemical losses in runoff from every field, we need to relate the transfer of chemicals from soil to runoff with soil properties and hydrological conditions, in order to identify the problem areas and devise methods for minimizing the losses.

Some relationships between soil properties and chemical transfer have been reported. Romkens and Nelson (1974) and Sharpley *et al.* (1978) reported a high linear correlation between the pre-rain-available phosphorus in surface soil and the dissolved inorganic phosphorus in runoff. McDowell *et al.* (1980) found that the equilibrium phosphorus concentration of soils (EPC) was related to the average concentration of dissolved phosphorus in runoff within a factor of 2 to 3. Leonard *et al.* (1979) found that for moderately adsorbed herbicides, the storm-average dissolved herbicide concentration in runoff from small watersheds was highly correlated with the prestorm herbicide concentration in 0- to 1-cm depth of soil, on a log–log basis. The data for seven different herbicides, four watersheds, and 4 yrs of study were fairly well described by a single relationship. For more strongly adsorbed herbicides, the total herbicide concentration in runoff was correlated with the product of herbicide concentration in soil and the sediment concentration in runoff. Ponce (1975) showed that the total dissolved solids in runoff from Mancos Shale lands were highly correlated with concentrations in the 0- to 0.25-cm depth of soil, but poorly with concentrations in the 0- to 3-cm depth interval, suggesting that a very thin zone of soil interacts with rainfall and runoff. Miller and Daniel (1981) related annual losses of soluble phosphorus in runoff from midwestern soils to rainfall, area, slope, land use, and farming practices on a regional scale. Along these lines, Haith and Tubbs (1981) proposed a loading function approach for a broad assessment of chemical losses in runoff. Wauchope and Leonard (1980) derived an empirical prediction equation for maximum pesticide concentration in runoff, within an order of magnitude, from application rate, pesticide formulation, method of application, and the time of first runoff event after application.

Several efforts have been made toward mathematically modeling the processes of chemical transport from land surfaces, in conjunction with existing models for water and sediment transport (Crawford and Donigian, 1973; Bruce *et al.*, 1975; Frere *et al.*, 1975; Donigian and Crawford, 1976; Donigian *et al.*, 1977; Williams and Haan, 1978; Knisel, 1980; Haith, 1980). Current versions of the two major efforts, which include both the nutrients and pesticides, are the ARM (Agricultural Runoff Management) model, developed for the U.S. Environmental Protection Agency by Hydrocomp, Inc. (Donigian *et al.*, 1977), and the CREAMS (Chemicals, Runoff and Erosion from Agricultural Management Systems) model, developed by a group of scientists in the U.S. Department of Agriculture (Knisel, 1980). Where possible, these models incorporate physically based descriptions of the various processes. However, because of the absence of the needed information, some of the mechanisms were oversimplified. Limited field tests of both the ARM and CREAMS models showed that the prediction of the chemicals transported entirely or partly in solution form was not satisfactory, particularly for individual runoff events (Donigian *et al.*, 1977; Davis and Donigian, 1979; Frere *et al.*, 1980; Leonard and Wauchope, 1980). These results indicate the need for a better understanding of the rainfall–runoff–soil interactions, the nature of the partitioning mechanisms between the solid- and solution-phase portions of an adsorbed chemical, and the solution flow pathways over and through the soil, in relation to surface runoff and shallow subsurface flow (interflow).

The purpose of this contribution is to analyze carefully the current state of the art in modeling the above processes and present new basic information to improve understanding and characterization of the phenomena. The discussion is devoted entirely to the transfer of soil-applied chemicals to runoff in solution phase. The transport of chemicals in solid phase is directly linked to soil erosion, for which a considerable amount of information is already available in the literature.

II. The Rainfall–Runoff–Soil Interaction

Chemicals at or near the soil surface can be transferred to overland flow in solution form by the mixing of rainwater with soil solution, dissolution of the chemical partly present in solid form, desorption of an adsorbed or absorbed chemical from the soil and crop residues in place, and desorption of the chemical from eroded sediment (Bailey *et al.*, 1974). The amount of chemical desorbed from eroded sediment is generally much less than the amount desorbed from soil in place. Furthermore, the chemical present on the sediment is likely to have been desorbed partially before the sediment is dislodged. Therefore, for simplicity the desorption from sediment may be considered as part of the desorption from soil in place. The mixing and desorption are accelerated by the kinetic energy of rainfall and runoff but also depend on the soil properties and soil conditions. The infiltration rate of the soil controls the rates of runoff and chemical transfer.

It has generally been assumed that a thin zone of soil at the surface undergoes mixing and desorption in relation to the transfer of chemicals to overland flow. Huff and Kruger (1967), in describing the transport of radioactive aerosols in a watershed, assumed that the effective soil mass interacting with rainwater was the mass of soil whose total particle surface area was equal to the land area, plus the mass suspended in flow. In their Agricultural Chemical Transport Model (ACTMO), Frere *et al.* (1975) assumed that the concentration of chemicals in runoff water was equal to that in the soil-pore water at the soil surface, which was calculated by using the theory of convective dispersion of chemical into the soil. Donigian *et al.* (1977) used a surface layer thickness of 0.2 to 0.6 cm based on some field calibrations of the ARM model. It was assumed that the rainfall mixed completely with the soil and soil water in this layer. The chemical in solution, as a result of this mixing, was apportioned among infiltration, runoff, and soil water in the layer in proportion to the amount of water in each mode. The CREAMS model incorporates essentially the same concept of a fixed mixing zone. The thickness of the mixing zone is defined as 1.0 cm, but it is assumed that only a fixed fraction of chemical available in this zone mixes with runoff water (Frere *et al.*, 1980; Leonard and Wauchope, 1980). The value of this fraction, designated as an extraction coefficient, may be equated with the thin zone assumed by Donigian *et al.* (1977). The values of this coefficient, or the thin mixing zone depth, for certain field data were between 0.01 and 0.4 (Frere *et al.*, 1980; Leonard and Wauchope, 1980). The concept of a thin mixing zone is illustrated in Figure 1.

Ahuja *et al.* (1981a) examined the depth of interaction between rainfall, overland flow, and soil, and the degree of interaction with depth by using ^{32}P (a relatively immobile isotope) as a tracer. In prewetted soil boxes with

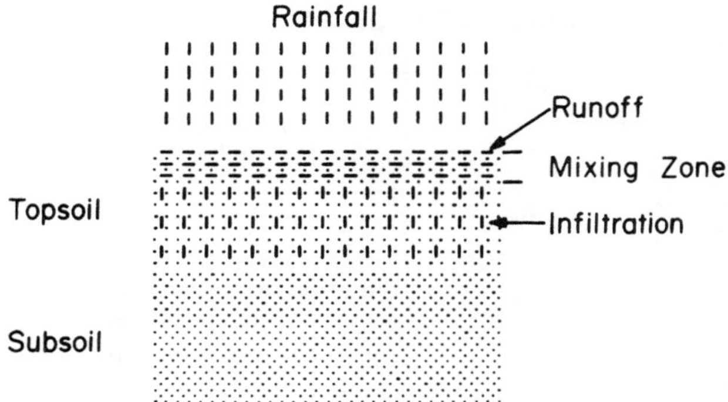

Figure 1. A schematic diagram illustrating the concept of a thin zone of complete mixing between rainwater and soil solution. Reproduced from Ahuja and Lehman, *J. Environ. Qual.*, Vol. 12, 1983, pp. 34–40 by permission of the Am. Soc. Agron., Crop Sci. Soc. Am., Soil Sci. Soc. Am.

pervious bottoms, a known amount of ^{32}P was placed at either 0.0-, 0.5-, 1.0-, 1.5-, or 2.0-cm depth. During a simulated rainfall, the runoff from each of these boxes was collected and analyzed for ^{32}P concentrations. The results showed that although some ^{32}P was released to runoff even from the 2.0-cm depth, the degree of interaction was maximum at the soil surface and decreased very rapidly with depth. The data for the Ruston soil, one of the three examined by Ahuja *et al.* (1981a), are graphed in Figure 2. The Ruston soil is a fine-loamy, siliceous, thermic Typic Paleudult. In reality, therefore, a thin zone of uniform interaction or mixing and fixed thickness does not exist. To determine if the model of Donigian *et al.* (1977) could be used as an approximation, an effective depth of interaction (EDI) was assumed to exist, within which the degree of interation (or mixing) was uniform and equal to that at the soil surface. The EDI was determined by taking the area under the curve of the rainstorm-average ^{32}P concentrations in runoff vs. the depth of ^{32}P placement, and dividing this area by ^{32}P concentration from soil surface (0.0 cm) placement of the tracer (illustrated in Figure 2). Effective average depth values for three different soils, on a 4% slope, were between 0.2 and 0.3 cm, when a 6.5 cm hr^{-1} simulated rainfall was applied for 30 min. The calculated EDI values were used in conjunction with the measured desorption kinetics of ^{32}P placed at the soil surface to predict the release in runoff of nonradioactive P that was mixed uniformly with the entire soil in the box. The predictions were good, and the concept of an effective mixing depth seemed adequate for P release under free infiltration conditions. This study also showed, however, that the EDI increased somewhat with time during a rainstorm, especially in early stages.

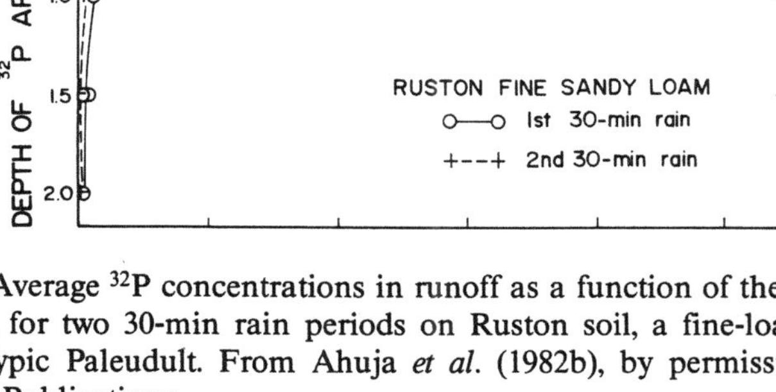

Figure 2. Average ^{32}P concentrations in runoff as a function of the depth of ^{32}P placement for two 30-min rain periods on Ruston soil, a fine-loamy silceous, thermic Typic Paleudult. From Ahuja *et al.* (1982b), by permission of Water Resources Publications.

The concept of EDI was later examined for the release of a non-adsorbed soluble chemical, Br, from soil boxes to runoff for one soil (Ahuja, 1982). Under free infiltration conditions, most of the Br in the mixing zone was lost by infiltration and runoff during the first 5 min. The concentrations of Br in runoff with time were within the bounds predicted by the model of Donigian *et al.* (1977) using EDI values at 2.5 and 10 min as determined by the ^{32}P method (Ahuja *et al.*, 1981a) for the same soil and rainfall conditions. Even though the assumed effective depth of interaction (EDI) increased with time, the use of a time-averaged EDI appeared adequate for practical predictions under free infiltration conditions.

A more critical test of the concept of an EDI within which the rainwater and soil solution mix completely during the transport process was provided by measuring the effects of varying infiltration rates on Br transfer to runoff (Ahuja and Lehman, 1983). A constant rate of simulated rainfall (6.8 cm hr^{-1}) was applied on boxes of three different soils that had been prewetted to near saturation with a bromide (4000 mg L^{-1} Br) solution. The bottom of the soil boxes was either an impervious plate that allowed no downward infiltration through the soil during the rainfall, a pervious plate that allowed free infiltration, or a pervious plate covered with a thin layer of soil slurry, which allowed infiltration at a reduced level. Bromide concentrations in runoff from impervious-base boxes were one to two orders of magnitude higher than in runoff from pervious-base boxes for all the soils (e.g., Figure 3). The concentrations in slurry-base boxes were intermediate. These results contradict the concept of EDI as shown below.

Assume a constant rate of rainfall (*R*) on a small area of saturated permeable soil with negligible surface depression storage or ponding. If the mixing of rainwater and soil water within the EDI is instantaneous and complete, then the mass balance of a water-soluble chemical in the EDI may be written as:

$$d(\mathrm{EDI} \cdot \theta \cdot C)/dt = -RC \qquad [1]$$

where *C* is the concentration of chemical in soil water, water percolating below the EDI, and the runoff water at any time *t*, and θ is the volumetric soil water content at saturation (Donigian *et al.*, 1977; Steenhuis and Walter, 1980). For equation [1], it is assumed that during the rainfall there is no transfer of chemical from below the EDI to the EDI by diffusion or other means. For constant EDI and θ, equation [1] is integrated to obtain:

$$C = C_0 \exp\left[-Rt/(\mathrm{EDI} \cdot \theta)\right] \qquad [2]$$

where C_0 is the initial concentration of the chemical in soil water. Thus, the logarithm of the runoff concentration is linear with time.

The assumption of complete mixing of rainwater and soil water within the EDI implies that the concentration of a mobile chemical in runoff

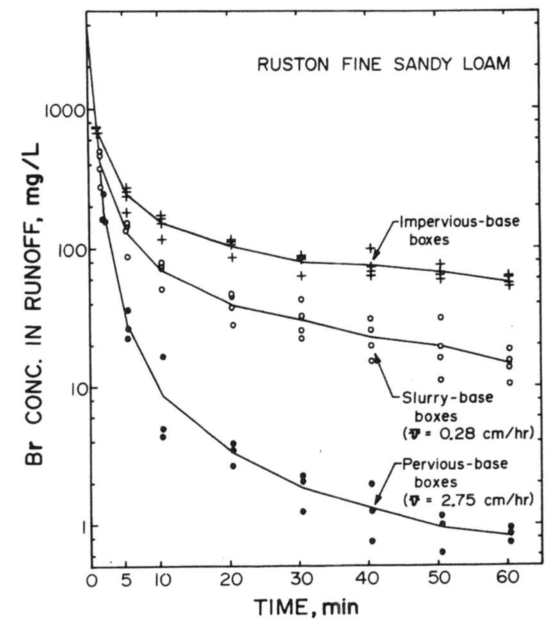

Figure 3. Bromide concentrations in runoff at different times during the rainfall in Ruston fine sandy loam boxes with an impervious base, a slurry base, or a pervious base. The curves are drawn through means of replicate data points. $\bar{v}$ is the average infiltration rate. Reproduced from Ahuja and Lehman, *J. Environ. Qual.*, Vol. 12, 1983, pp. 34–40, by permission of the Am. Soc. Agron., Crop Sci. Soc. Am., Soil Sci. Soc. Am.

water or water infiltrating below the EDI will be equal to that in soil water within the EDI at all times. For a given rainfall rate R, the concentrations in runoff described in equation [2] are the same irrespective of whether there is downward infiltration into the soil or not. To the contrary, the large effects of infiltration rates noted in the above experimental results and nonlinearity of the concentration curves on the semilogarithmic plots (Figure 3) indicated that the concept of an effective depth with complete mixing of rainwater and soil solution was not strictly valid.

Experimental measurements of Ingram and Woolhiser (1980) in impervious-base soil boxes indicated that near the end of rainfall events the concentration of a soluble chemical in soil water in a thin surface zone was much higher than that in runoff water. Based on this, they suggested that runoff water might not mix completely with soil water within the mixing zone. If it is assumed that the degree of mixing of runoff water with soil water within the EDI is λ, and that of the water infiltrating through the EDI is γ, then the mass balance equation is:

$$d(\mathrm{EDI} \cdot \theta \cdot C)/dt = -\gamma(R - Q)C - \lambda QC \qquad [3]$$

where Q is a constant runoff rate. This specifies the concentration of chemical in runoff to be λC at any moment of time, while the concentration in soil water within the EDI is C. Equation [3] can be integrated to obtain the concentration in runoff:

$$\lambda C = \lambda C_0 \exp\left[\frac{-[\gamma R - (\gamma - \lambda)Q]}{\mathrm{EDI} \cdot \theta} t\right] \qquad [4]$$

This relationship indicates that the logarithm of the concentration in runoff will still be a linear function of time for constant λ, γ, EDI, R, and the infiltration rate, $R - Q$ (including no infiltration, when $Q = R$). For a fixed R but different Q values, the intercept of this semilogarithmic relationship will be the same [$= \log(\lambda C_0)$], whereas the slope will be smaller for larger Q values. The data of Figure 3 indicate that equation [4] does not really hold either.

The results in Figure 3 and the Br concentrations in soil water with depth in impervious-base soil boxes measured at the end of 1 hr of rainfall events (Figure 4) showed that Br was released to runoff from as great as 2.0-cm depth, but the amount released decreased exponentially with depth. The concentration of Br in soil water within the top 0.25 cm layer of soil was higher than in the runoff near the end of rainfall event, as also reported by Ingram and Woolhiser (1980). This apparently results from the sharp increase in the Br concentration in soil from 0.0- to 0.25-cm depth (Figure 4). The above results agree with those of ^{32}P study (Ahuja *et al.*, 1981a) that the degree of interaction or mixing between runoff and soil water is nonuniform and decreases rapidly with depth below the soil surface.

Snyder and Woolhiser (1985) studied the transfer of a dye from a saturated sand box to overland flow with and without infiltration. With no infiltration, the shape of their dye concentration profiles or the dye removal with depth (Figure 5a) is very similar to that of Br (Figure 4). However, the degree of interaction for the dye does not decrease as rapidly with depth as for Br. The Ottawa sand used by Snyder and Woolhiser had an extremely high saturated hydraulic conductivity (260 cm/hr^{-1}), which seemed to increase the degree of interaction (or mixing) with depth. With infiltration, the dye concentration profiles were S shaped, because of the displacement caused by downward movement (Figure 5b). These results also suggest that whereas for low infiltration rates the concept of EDI and complete mixing is not suitable, the concept may be an adequate approximation where high infiltration rates occur.

The transfer of chemical from below the soil surface to the surface is a process of turbulent mixing and dispersion caused by raindrop impact. The process results in a nonuniform degree of mixing with depth between rainwater and soil solution. A mixing model that incorporates this nonuniformity with depth is described in a later section. The above mechanism of transfer may also be considered an accelerated diffusion

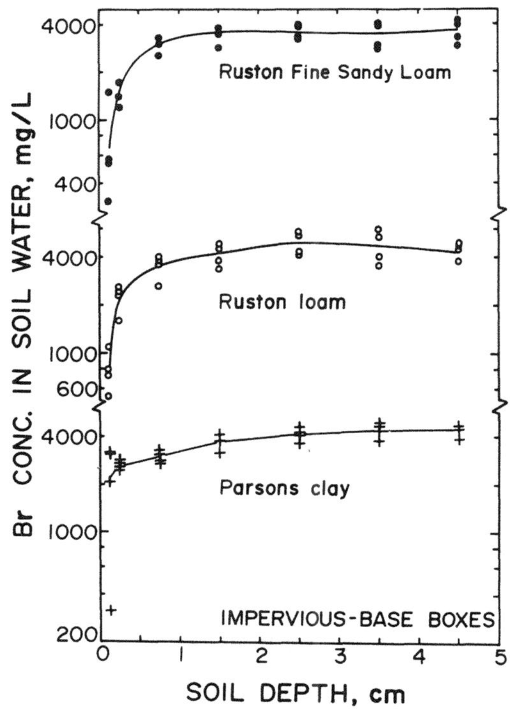

Figure 4. Bromide concentrations in soil water with depth in soil boxes of the three soils with impervious bases at the end of 1-hr rainfall events. The Ruston soils are fine-loamy, siliceous, thermic Typic Paleudult, and Parsons clay is a fine, mixed, thermic Mollic Albaqualf. Reproduced from Ahuja and Lehman, *J. Environ. Qual.*, Vol. 12, 1983, pp. 34–40, by permission of the Am. Soc. Agron., Crop Sci. Soc. Am., Soil Sci. Soc. Am.

process. To describe this transfer in the presence of infiltration, the conventional convection–diffusion equation (Nielsen and Biggar, 1963) may be adopted:

$$\frac{\partial C}{\partial t} = \frac{\partial}{\partial z}\left[T\frac{\partial C}{\partial z} - VC\right] \qquad [5]$$

where C is the concentraiton of the chemical in soil water at a certain time t, z is the soil depth, V is the flow velocity, and T is the transfer coefficient. Runoff concentration equals C at $z = 0$. Equation [5] will need to be solved subject to known initial conditions, known rates of rainfall and runoff (R and Q), and setting the flux of chemical at the soil

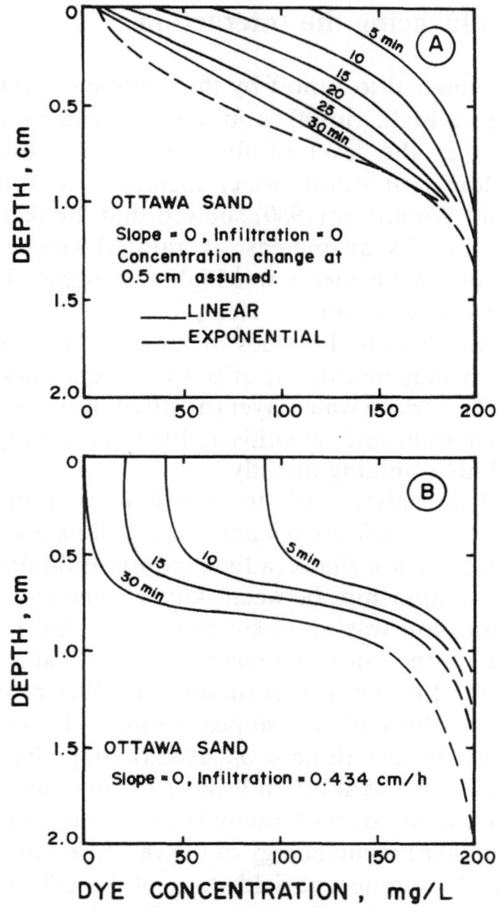

Figure 5.(A) Average dye concentration profiles in Ottawa sand boxes at slope = 0 and infiltration rate = 0, during a simulated rainfall. From Snyder and Woolhiser (1985), by permission of the authors and the American Society of Agricultural Engineers. **(B)** Average dye concentration profiles in Ottawa sand boxes at slope = 0 and infiltration rate = 0.434 cm hr^{-1} during a simulated rainfall. From Snyder and Woolhiser (1985), by permission of the authors and the American Society of Agricultural Engineers.

surface equal to $QC_{z=0}$. For these conditions, an analytical solution exists in the literature for V = 0 (Crank, 1964, p. 34). This solution was tested on the impervious-base data of Figure 3 (L. R. Ahuja, 1984, unpublished), but was found to be unacceptable. Perhaps, the transfer coefficient T should not be a constant but a decreasing function of depth. This provision will make the convection–diffusion approach more complicated in parameter evaluation and solution.

III. Factors Influencing the Interaction

The transfer of chemical to runoff by the turbulent mixing mechanism is influenced by rainfall, runoff, and soil characteristics. Wiese *et al.* (1980) reported that the level of fluometuron [1,1-dimethyl-3-(*a,a,a,*-trifluoro-*m*-tolyl)urea] in runoff water increased with increase in soil slope. Ingram and Woolhiser (1980) showed that the release of $CaS0_4$ to runoff was enhanced by an increase in rainfall kinetic energy and in slope. The effect of kinetic energy was higher at higher slopes. The effect of rainfall intensity was not as clear, but an increase in intensity enhanced the transfer at high values of kinetic energy and slope. Both slope and intensity may modify the effects of kinetic energy by changing the thickness of the surface water layer (depth of overland flow) prevalent during rainfall (Ingram and Woolhiser, 1980). However, they may also increase the turbulent mixing directly.

The release of phosphorus (P) to runoff in solution form was also enhanced by rainfall kinetic energy and slope (Ahuja *et al.*, 1982a). In two soils, Ruston and Houston Black (a fine, montmorillonitic, thermic Udic Pellustert), the relationship between kinetic energy and average P concentration in runoff was approximately linear for a given soil slope (Figure 6). However, the rate of increase was greater at a higher slope, in agreement with the finding of Ingram and Woolhiser (1980). At a fixed kinetic energy, the effect of soil slopes, between 4 and 16%, on the P release was nearly linear (Ahuja *et al.*, 1982a). Slope length, between 33 and 100 cm, also increased solution P release. This was explained by the effect of increased water to soil ratio on the P desorption (Sharpley *et al.*, 1981a), but the higher kinetic energy of the increased overland flow may also be involved. A more recent field study of the effects of slope length on P, K (potassium), and Br release to runoff indicated that these effects, if any, were small relative to the natural spatial variability of concentrations (Lehman and Ahuja, 1985a).

Hydraulic conductivity of the soil surface layer can also affect chemical transfer by influencing the depth and degree of mixing, especially if the subsurface hydraulic conductivity is low (Sharpley *et al.*, 1981b; Ahuja and Lehman, 1983). This may be important in a freshly tilled soil, when the hydraulic conductivity of the surface layer is very high and the chemical is mixed in the tilled layer. Soil aggregates and clods associated with tillage, which tend to retain the chemical near the surface by reducing its leaching, further enhance the effects of hydraulic conductivity. Ahuja *et al.* (1983) found that the transfer of Br to runoff was 1 to 2 orders of magnitude greater from a cloddy surface layer than from a sieved soil layer.

As a first approximation, we may assume that the soil, slope, rainfall, and runoff characteristics influence the chemical transfer process in a manner similar to their effect on soil erosion. The limited experimental

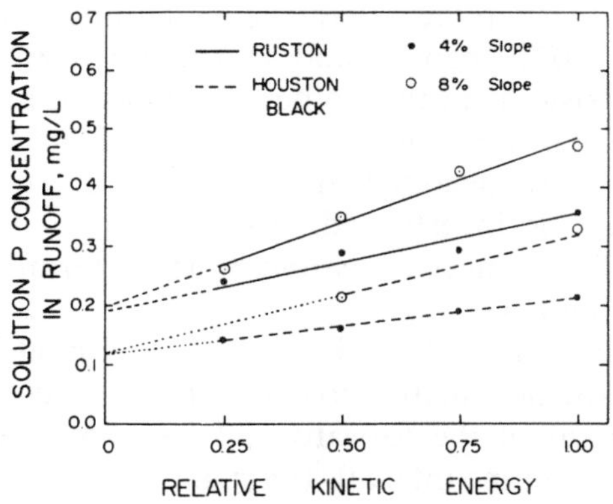

Figure 6. Solution P concentration in runoff as a function of relative kinetic energy of the simulated rainfall for Ruston fine sandy loam and Houston Black clay soils at 4 and 8% slopes. The average values are means of two 30-min rainfalls. From Ahuja *et al.* (1982b), by permission of Water Resources Publications.

results cited above do seem to agree with the following conceptual model, analogous to the physically based soil erosion equation (Foster *et al.*, 1977):

$$M = (aS + b)E + cL^mS^eq^n \quad [6]$$

where M is the mass of chemical transferred to runoff, S is the slope, E is the kinetic energy per unit volume of rainfall, L is the slope length, q is the runoff rate per unit area, and $a, b, c, e, m,$ and n are constants for a given soil. The first term on the right-hand side of equation [6] represents the effect of raindrop impact, and the second term is the effect of overland flow. A great deal of field experimentation is needed to determine the relative importance of each term and the values of the parameters.

IV. The Desorption of Adsorbed Chemicals

In the ARM model, a nonlinear (Freundlich-type) equilibrium isotherm is used as a basis for partitioning the labile portion of pesticides between adsorbed and solution phases (Donigian and Crawford, 1976). The fact that some of the pesticide is in a fixed, unavailable state, is recognized. Hysteresis in the equilibrium isotherm can also be included. In the CREAMS model, a constant partition coefficient, i.e., an equilibrium proportional isotherm, is assumed between the two phases (Leonard and Wauchope, 1980). Field tests of both these models have shown poor predictions for solution-phase pesticides in runoff, especially on an individual-event basis rather than on a monthly or annual basis (Donigian *et al.*, 1977; Leonard and Wauchope, 1980). Values for the proportionality constants in the isotherms seemed to increase with time during the course of some rainfall events (Leonard and Wauchope, 1980). Donigian *et al.* (1977) suggested that perhaps the use of some kinetic approaches should be investigated. However, the above discrepancies may be a result of misconceived assumptions concerning the nature and extent of the mixing zone.

For P and other adsorbed nutrients, the ARM model incorporates a reversible linear kinetic equation to calculate desorption and transfer to runoff. Computation of the forward and backward reactions is done numerically for small time intervals (Donigian and Crawford, 1976). The reaction constants are assumed to be independent of the water to soil ratio. In the CREAMS model, the kinetics of the P release is not invoked, per se. It is assumed that the soil contains a certain amount of readily soluble P, and that the buffering capacity of the soil for P maintains a certain fixed concentration in runoff, percolating water, and soil water (Frere *et al.*, 1980). In limited field tests, the ARM model seemed to give reasonable calculations of the available P in soil at different times and for monthly amounts in runoff, but solution P concentrations on a runoff event basis were not predicted well (Davis and Donigian, 1979).

Travis and Etnier (1981) surveyed the relationships in current use to quantify sorption–desorption of reactive solutes by soil, covering both the equilibrium and kinetic models. The equilibrium models include the linear, Freundlich, Langmuir, two-surface Langmuir, and competitive Langmuir isotherms. The kinetic models include the reversible linear, reversible nonlinear, kinetic product, bilinear, diffusive mass transfer, Elovich, Faya and Eyring, and two-site kinetic models. The soil is an extremely complex heterogenous system. The adsorption mechanisms include weak adherance to the surfaces, strong and specific chemisorption, and ion exchange. The equilibrium isotherms and kinetic models are used to represent the totality of these processes on a variety of soil components. No one model, therefore, may be suitable for all soils and conditions. The choice also depends on the problem, type of knowledge available, and accuracy required (Travis and Etnier, 1981). Many situations lend themselves to analysis with simpler equilibrium models.

Most of the work in the past has been done to define transformations and changes in soil nutrients (especially P) available for plant growth, involving long periods of time. Since the time the environmental quality problems were recognized, there have been many studies on the movement of pesticides, nutrients, and other solutes through soils, which require short-term adsorption–desorption behavior of these chemicals (see Travis and Etnier, 1981, for a comprehensive review). However, most of these studies have involved testing the applicability of one or more of the models described above. Very few studies have been conducted on the actual kinetics of adsorption–desorption within the first few minutes, which is important for the chemical transfer to runoff. Lindstrom *et al.* (1970), Wahid and Sethunathan (1978), and Wauchope and Myers (1985) have reported such studies with some common pesticides; these indicate that the adsorption or desorption reaction can approach 90 to 95% equilibrium within minutes, but that achieving actual equilibrium can take hours. For atrazine [2-chloro-4-(ethylamino)-6-(isopropylamino)-*s*-triazine] and linuron [3-(3,4-dichlorophenyl)-1-methoxy-1-methylurea], 50% of the reaction was completed in less than 2 min, and 75% in 3 to 6 min (Wauchope and Myers, 1985). Within the error of measurement, some of the experimental data of Wauchope and Myers (1985) suggested almost instantaneous equilibrium. Leenheer and Ahlrichs (1971) found that 60% desorption of carbaryl (1-naphthyl-*N*-methylcarbamate) and parathion [*0,0*-diethyl-*0*-(*p*-nitrophenyl)phosphorothioate] on organic matter was achieved in 1 min. Based on such results and determinations of activation energy of the reactions, several workers have suggested that diffusion is generally the rate-limiting factor for adsorption and desorption of weakly to moderately adsorbed pesticides in soils (Haque and Sexton, 1968; Haque *et al.*, 1968; Lindstrom *et al.*, 1970; Leenheer and Ahlrichs, 1971). However, a variety of kinetic models, such as the reversible linear (Lindstrom *et al.*, 1970), the first-order nonlinear Faya and Eyring (Haque and Sexton, 1968; Leenheer and Ahlrichs, 1971), and two-sequential-linear reactions model (Wauchope and Myers, 1985), can be used to describe the results. Simple proportional relationships may be adequate for moderately adsorbed pesticides in practical applications.

For soil P, there is a wide range of activation energies for desorption (Posner and Bowden, 1980) and, therefore, the P exists in several forms—labile, moderately fixed, and strongly fixed. Several studies have indicated that the desorption of soil P during a short period of time (labile P) is a low activation energy process and proposed that it is diffusion limited (Kuo and Lotse, 1974; Evans and Jurinak, 1976; Cooke, 1966; Enfield and Bledsoe, 1975; Vig and Dev, 1979). The linear reversible kinetic model does not adequately describe P desorption for intermediate times (Amer *et al.*, 1955; Evans and Jurinak, 1976). Several nonlinear models have been found adequate (Sharpley *et al.*, 1981a), but a simple equation involving a power-form time dependence for desorption has proved quite successful in several cases (Kuo and Lotse, 1974; Barrow, 1979; Chien and Clayton, 1980). Sharpley *et al.* (1981a) found this equation applicable in five different soils at various levels of P additions for times up to 180 min. The effect of water to soil ratio on desorption was significant, but it could be described by a factor consisting of this ratio raised to a power. The results for two of the soils, a Houston Black clay and Kirkland silt loam (a fine, mixed, thermic Udertic Paleustoll), are summarized in Figure 7. Sharpley *et al.* (1981b) and Ahuja *et al.* (1982b) related the laboratory-determined kinetic equations to the chemical transfer to runoff from soil

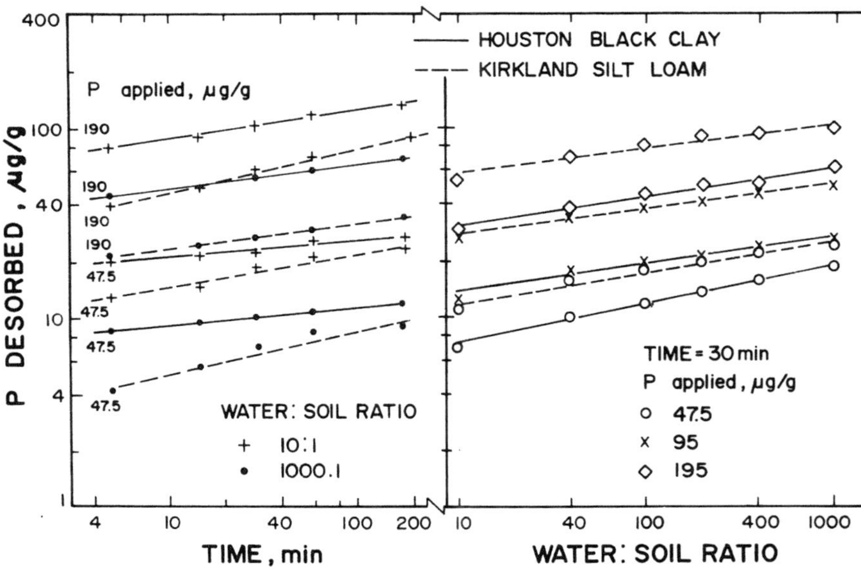

Figure 7. Log–log plot of P desorbed as a function of time at a given water/soil ratio, and as a function of water/soil ratio at a given time, for two different soils. From Ahuja *et al.* (1982b), based on the work of Sharpley *et al.* (1981a), by permission of Water Resources Publications.

boxes under simulated rainfall. Sharpley *et al.* (1982) also used the laboratory parameters to predict solution P concentrations in runoff from small watersheds with fair success.

The desorption characteristics of a soil may best be measured during an actual rainfall, because the rainfall–runoff–soil interaction involves a different kind of mixing than the laboratory batch technique of measurement. Some of the kinetic or equilibrium parameters may not be the same in the two cases (Sharpley *et al.*, 1981b; Ahuja *et al.*, 1982b). However, for measurements taken during an actual rain, complications caused by the variable degree of rainwater–soil interaction with depth need to be taken into account.

Besides the adsorption mechanisms discussed above, another mechanism that may substantially influence the desorption of a chemical from soil during a rainfall is the physical absorption and retention of the chemical within soil aggregates and clods. The data in Figure 8 show that the concentrations of Br (a nonadsorbed chemical) in runoff from a surface layer of small clods or aggregates containing Br were 1 to 2 orders of magnitude greater than that from the same soils without clods, even though the infiltration was 2 to 3 times greater in the former (Ahuja *et al.*, 1983). In another experiment, it was found that the grass cover in a field plot also tremendously increased Br transfer to runoff as compared with the transfer from the same soil with no grass in the laboratory (Heathman

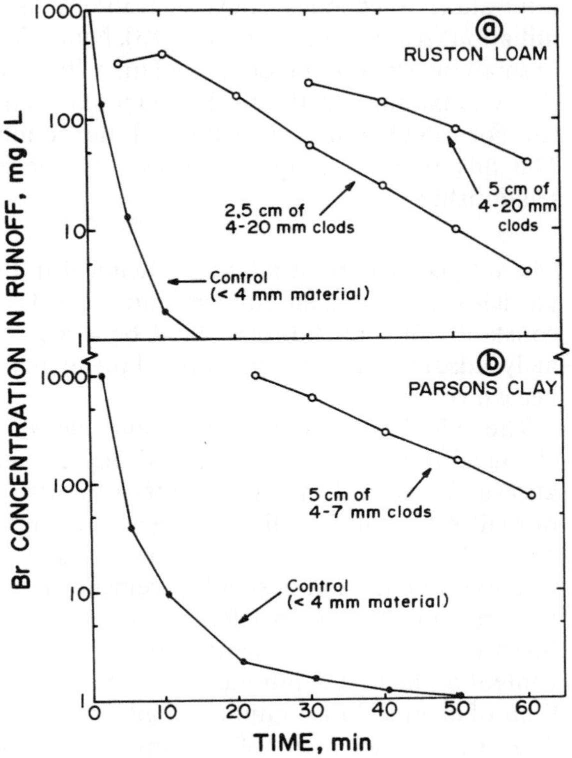

Figure 8. Bromide concentrations in runoff from clod and control boxes of: (a) Ruston soil; (b) Parsons soil. Reproduced from Ahuja *et al.*, *Soil Sci. Soc. Am. J.*, Vol. 47, 1983, pp. 746–748, by permission of the Soil Science Society of America.

et al., 1985a). Ingram (1976) also reported higher Br in runoff from a grass plot than from a fallow plot. Apparently, the grass thatch and foliage were absorbing and retaining Br, which decreased the infiltration of this chemical into the soil but increased its transfer to runoff. Crop residue on the soil surface may also shield some of the chemical from mixing with rainwater. These types of physical absorption and retention processes cannot be easily distinguished from the chemical adsorption processes. The presence of physical absorption makes it necessary to measure the transfer properties under actual field conditions. The desorption of plant nutrients from plant residues left on the soil surface in substantial amounts, and even from the living plants, may need to be characterized separately (Schreiber, 1985; Schreiber and McDowell, 1985; Sharpley, 1981).

V. Modeling the Transfer to Runoff

A. A Nonuniform Mixing Model

The model incorporates the varying degree of mixing between rainwater and soil water during chemical transfer to runoff, as well as the effects of infiltration on chemical movement into the soil before and after runoff initiation (Ahuja and Lehman, 1983). Numerical computations are made in small intervals of soil depth and time. Based on the evidence in Section II, the maximum depth of soil interacting with rainfall is taken to be 2.0 cm. For calculations, this is divided into twenty 1.0-mm depth intervals. The adsorption-desorption process is represented by a proportional relationship:

$$C_s = \alpha C \qquad [7]$$

where C_s is the concentration of chemical in the adsorbed phase on soil particles (g g^{-1}), C is the concentration in soil solution (g mL^{-1}), and α is a constant. This relationship may be adequate for weakly to moderately adsorbed chemicals. Other kinetic relationships can be used if necessary.

The initial soil water content, bulk density, initial concentration of chemical in solution and adsorbed phases, and value of α are assumed known for each 1-mm depth interval. Also assumed known are the rainfall rate, runoff initiation time, and runoff rates. Before the runoff starts, all the rainwater infiltrates into the soil. For calculations, this water is moved into the soil in small increments. Each increment is set equal to the amount of water needed to saturate the soil in one 1-mm depth interval. The first increment of the water is assumed to raise the water content of the first depth interval to saturation and dilute the concentrations of chemical in solution and adsorbed phases, which are calculated. The second increment of the infiltrating water first mixes with the solution and soil in the first depth interval, and then an equivalent amount of the resulting solution moves downward to saturate and mix with the second depth interval. The subsequent increments of infiltrating water are routed similarly to successive depth intervals, until the entire amount of rainwater, prior to runoff initiation, has been routed. Generally, all the twenty 1-mm depth intervals are saturated before this time.

Starting from the time of runoff initiation, it is assumed, based on evidence presented in Section II, that the degree of mixing between the rainfall and soil solution, β, decreases exponentially with soil depth, as (Ahuja and Lehman, 1983):

$$\beta = \exp(-Bz) \qquad [8]$$

where B is a constant and z is the soil depth. Only at the soil surface, $z = 0$, is β equal to 1.0, implying complete mixing between rainwater and soil water. The value of constant B is known from a prior calibration, and an average β value is calculated for each 1-mm depth interval. A small increment of rainwater is mixed with solution and soil in each depth interval in proportion to its degree of mixing. The change in concentration of the chemical in any depth interval, i, is calculated by the mass balance equation for a unit area:

$$C_i^1(\theta_i + \alpha S_i) = C_i^2(\theta_i + \alpha S_i + \beta_i R \Delta t) \quad [9]$$

where C_i^1 is the concentration of chemical in solution in depth interval i before mixing with an $R\Delta t$ increment of rainfall (R is the rainfall rate and Δt is a small time step), C_i^2 is the concentration after mixing θ_i is the amount of water in the depth interval, S_i is the mass of soil in this interval, and β_i is its degree of mixing. After mixing with all the depth intervals, the concentration of chemical in $R\Delta t$ amount of rainwater, C_r^2 is given by:

$$C_r^2 = \left[\sum_{i=1}^{20} \beta_i C_i^2 R\Delta t\right] / R\Delta t = \sum_{i=1}^{20} \beta_i C_i^2 \quad [10]$$

A known portion of this water with concentration C_r^2 becomes runoff at this time, while the rest infiltrates into the soil. The portion that infiltrates is made to displace the soil solution in all depth intervals downwards, in the manner of a piston displacement. However, the time step Δt is kept small enough that the amount of water that infiltrates in this time step is less than θ_i (equal to about $\frac{1}{2}\theta_i$). This allows some mixing between adjoining depth intervals to take place during displacement as the average concentration is calculated in each depth interval after each time step. The above procedure is repeated until the concentrations in runoff with time are obtained for the rainfall period of interest.

The above simplified approach was applied to the data in Figure 3. The constant B was obtained by an optimization on the data for the impervious-base boxes that gave the best fit to the plot of log C vs. time. This value of B and the corresponding β values obtained from equation [8] were then used in calculations for the infiltration cases of slurry-base and pervious-base boxes of that soil. The results of the calculations are presented in Figure 9. The B value for this soil was 0.9. The results for two other soils were similar. There are appreciable deviations between the experimental and calculated curves at low concentrations. However, the major effects of infiltration rates are adequately represented. The deviations may result from such factors as retention of chemical within soil aggregates (Ahuja *et al.*, 1983) that are not included in the model. The method could possibly be improved somewhat by taking a function other than the exponential in equation [8], but this may require additional unknown parameters.

The predictions of the model for a grassed field plot, where the grass cover and thatch seemed to adsorb and retain Br, are shown in Figure 10 (Heathman *et al.*, 1985a). The α value was equal to 2.0 and B equal to 0.5.

Recently, the model was applied to the field-plot data of Baker and Laflen (1982) and Baker *et al.* (1982) from Iowa for four pesticides and three plant nutrients, NO_3, NH_4, and PO_4, applied on bare plots or on plots with varying amounts of plant residue on them (Heathman *et al.*, 1985b). In all cases, the model gave a reasonably good fit to the data. The

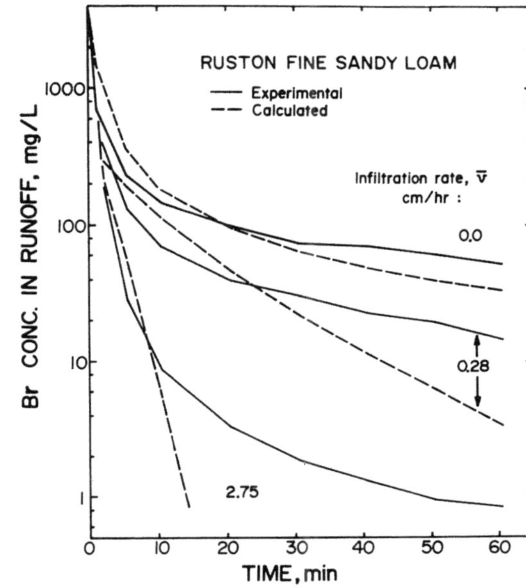

Figure 9. Bromide concentrations in runoff calculated by using an exponentially variable degree of mixing with depth below the soil surface, compared with the experimental data for Ruston fine sandy loam for the three infiltration rates. Reproduced from Ahuja and Lehman, *J. Environ. Qual.*, Vol. 12, 1983, pp. 34–40, by permission of the Am. Soc. Agron., Crop Sci. Soc. Am., Soil Sci. Soc. Am.

best-fit values of the parameters α and B were influenced to some extent by spatial variability of soil conditions within the replicated plots. The residue cover greatly increased the α value and to some extent the B value, suggesting that the cover was shielding some of the chemical on the soil surface from mixing with rainfall (an effect similar to absorption in aggregates) and also decreasing the degree of mixing with depth (by reducing the raindrop kinetic energy). The value of α was very high for P. Examples of model predictions for atrazine in runoff are shown in Figures 11 and 12, and for NH_4 and P in Figure 13. It is interesting that the model worked well in all cases, including P.

B. An Empirical Kinetic Model

Sharpley *et al.* (1981a) found the following empirical equation to provide a reasonably good description of P desorption from five soils in laboratory batch experiments:

$$P_d = AP_0 t^k W^p \quad [11]$$

where P_d is the amount of P desorbed in time t, P_0 is the initial desorbable P content of soil, W is the water to soil ratio, and A, k, and p are constants for a given soil. Assuming that this equation is also valid for chemical

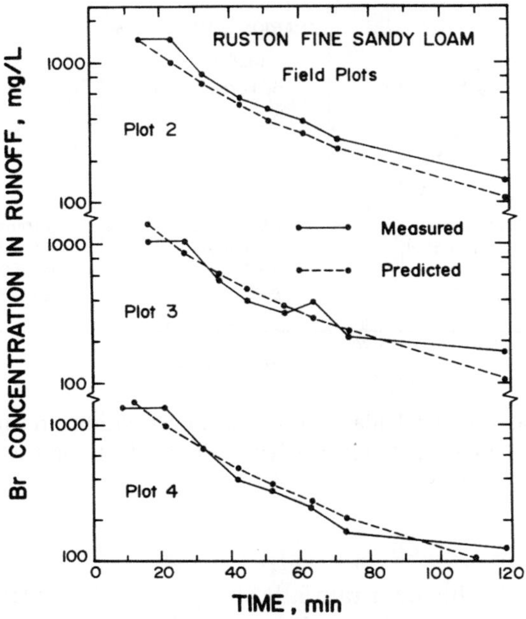

Figure 10. Predicted Br concentrations in runoff for field plots 2 to 4, Run No. 1, using the optimum values of B and α found for plot 1, compared with the measured values. From Heathman *et al.* (1985a), by permission of the authors and the American Society of Agricultural Engineers.

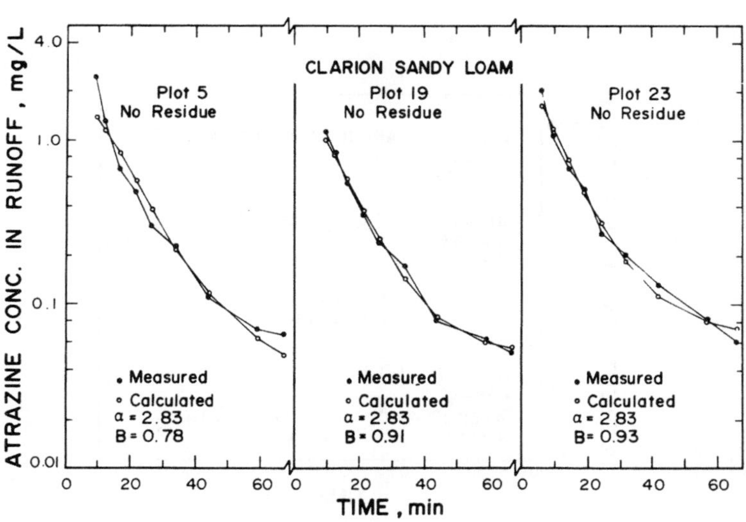

Figure 11. Measured and calculated atrazine concentrations in runoff from Iowa field plots with no plant residues on the soil surface. From Heathman *et al.* (1985b), by permission of the authors.

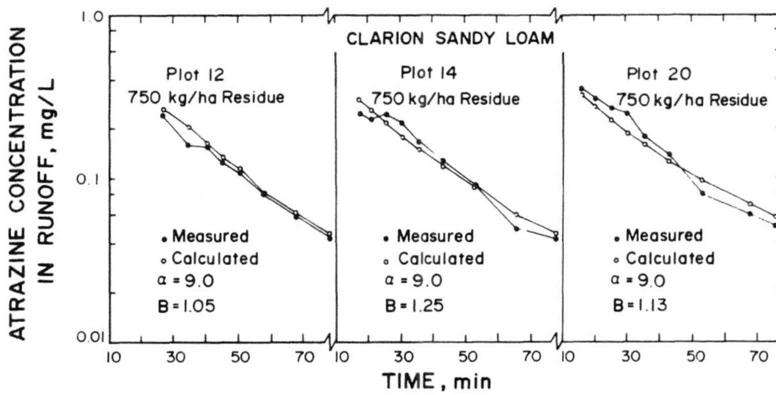

Figure 12. Measured and calculated atrazine concentrations in runoff from Iowa field plots at the 750-kg ha^{-1} residue level. From Heathman *et al.* (1985b), by permission of the authors.

transfer to runoff, during a rainfall, Sharpley *et al.* (1981b) obtained the expression for concentration of P in runoff water, C_{ro}, as:

$$C_{ro} = \frac{1}{R}\frac{dP_d}{dt} = AP_0kt^{k-1}W^P/R \qquad [12]$$

Equation [12] adequately described the experimental data for C_{ro} from soil boxes of the five soils under simulated rainfall. However, the

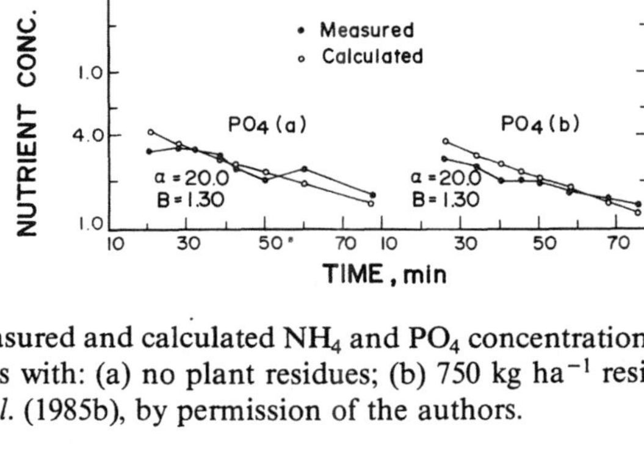

Figure 13. Measured and calculated NH_4 and PO_4 concentrations in runoff from Iowa field plots with: (a) no plant residues; (b) 750 kg ha^{-1} residue level. From Heathman *et al.* (1985b), by permission of the authors.

constants in the equation for C_{ro} were not the same as those derived from the batch studies.

Observations by the author over the years indicate that an equation of the form [12] may also be applicable for pesticides and other nutrients in runoff, including mobile and adsorbed or absorbed chemicals. For rainfall conditions, the water to soil ratio, W, may be replaced by the rainfall intensity, R. The term R^p may then be assumed also to include the effect of rainfall kinetic energy, related to intensity, on the turbulent transfer of chemicals. Based on the evidence in Section III, the effect of soil slope may be approximated as linear. With these modifications and replacing P_0 by M_0 to denote the initial desorbable mass of any chemical per unit weight of soil, equation [12] becomes:

$$C_{ro} = AM_0kt^{k-1}R^{p-1}S \quad [13]$$

On long slopes, the overland flow energy may also influence the chemical transfer. This factor is not included at this stage. Further work is needed to determine its importance under field conditions (Lehman and Ahuja, 1985a). The effects of varying infiltration rates, and hence of initial soil moisture conditions and infiltration before runoff initiation are also not included in equation [13].

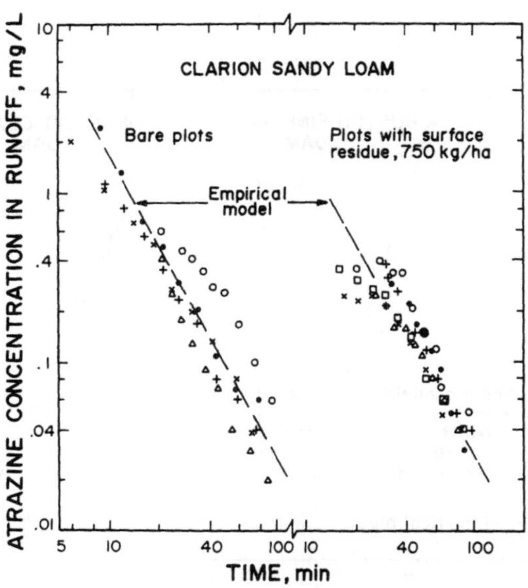

Figure 14. Log-log graphs of atrazine concentrations in runoff vs. time from replicated bare plots and plots with a residue cover on Clarion sandy loam soil. The dashed lines are eyeball fits of equation [13] to the data (From the data of Baker *et al.*, 1982).

A test of equation [13] for atrazine concentrations in runoff from six replicated bare and medium-residue-level field plots (Baker *et al.*, 1982) is provided by a log–log plot shown in Figure 14. The differences between replicates is a result of spatial variability of soil conditions, including initial soil moisture. The data for any given plot approximately fall along a straight line, in accordance with equation [13]. It appears that the effects of natural spatial variabilities on the results may be much more than the errors in the model. The data for another set of plots which had 750 kg ha^{-1} of residue cover on them are within the spatial variability of the data for bare plots. These results indicate that the effects of residue cover in reducing the total chemical load in runoff was mainly a result of increased infiltration and delayed runoff. However, the parameters of equation [13] are not appreciably influenced by the residue cover.

The tests of the model and the effects of initial soil moisture conditions on the Br and K in runoff from soil boxes (Heathman *et al.*, 1985a) are shown in Figures 15 and 16, respectively. The mobile chemical, Br, is very sensitive to the initial moisture and soil conditions, whereas an adsorbed chemical, K, is less sensitive. However, the model fit, shown for the wet initial condition, is fairly good in each case. The model also provided a good fit to the data of Figure 8 for Br transfer from small clods.

The experimental results of Schreiber and McDowell (1985) and Schreiber (1985) indicate that the desorption of nutrients from crop residues by a simulated rainfall could also be described by a power-form

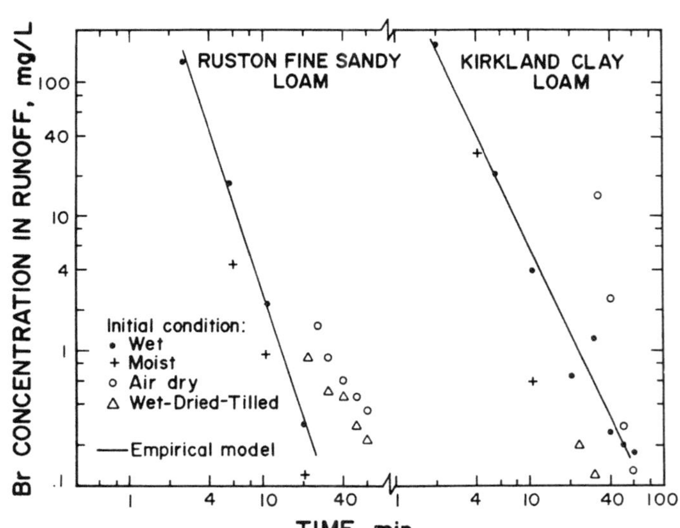

Figure 15. Log–log graphs of Br concentrations in runoff vs. time from two soils with different initial moisture conditions. The lines are drawn through the data for wet initial condition (From the data of Heathman *et al.*, 1985a).

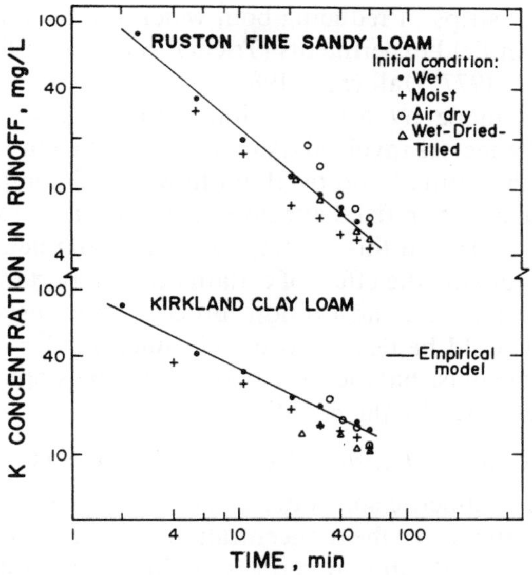

Figure 16. Log–log graphs of K concentrations in runoff vs. time from two soils with different initial moisture conditions. The lines are drawn through the data for wet initial condition (From the data of Heathman *et al.*, 1985a).

model, such as equation [13], both for concentration as a function of time and the effect of water to residue ratio (residue loading rate).

VI. Transport from Partial Land-Area Applications—Buffer Strips

Under certain situations in agriculture, the chemicals may be applied on only a portion of a sloping field or watershed. If only the upper part of a field is treated, its runoff containing the chemical will be "diluted" as it flows over and interacts with untreated soil and water before entering an outlet. If the chemical is applied only on the lower area, runoff from the untreated upper area will flow over and change the chemical concentrations of soil and water in the lower area. Relative lengths of the upper and lower areas, runoff rates, depth of soil interacting with rain and runoff, and the chemical adsorption–desorption properties of soil are factors that influence the concentrations of chemicals in runoff at the outlet. A better understanding of the processes and concepts pertaining to partial-area applications should help reduce the losses of chemicals from the field and aid in developing criteria for buffer strips at the lower end of a slope to filter out chemicals that are otherwise lost to runoff. Beneficial effects of buffer strips in reducing both water and chemical losses have been reported in field experiments (Trichell *et al.*, 1968; Baur *et al.*, 1972; Asmussen *et al.*, 1977; Hall *et al.*, 1983).

Calculations made by using a simple theory gave some valuable insights into the factors involved (Ahuja *et al.*, 1985). For simplicity it was assumed that the rainfall and overland flow mixed completely with soil solution within a certain fixed effective depth of soil, h. Even though it is shown in Section II that this assumption is not completely valid, this is used here for studying the effect of certain other important factors. It was also assumed that the adsorption process was instantaneous and reversible and could be described by the function $C_s = \alpha C$. With these assumptions, the mass balance at a point along the slope (x) and a given time (t) are described by the equation:

$$h[\theta + D\alpha]\partial C/\partial t = -\partial(qxC)/\partial x - vC \qquad [14]$$

where θ is the volumetric soil water content within the depth h; D is the soil bulk density; C is the concentration of chemical in soil water, overland flow, or infiltrating water; v is the infiltration rate per unit area; and q is the runoff rate. The parameters v and q can be treated as time dependent or taken as piecewise constant over small time intervals. These parameters are considered uniform over space, and the overland water movement is assumed to be a uniform sheet flow. A case of partial sheet flow is considered later. Equation [14] also assumes no surface detention storage or ponding. However, if necessary, the depth of ponded water can be added to the term multiplying $\partial C/\partial t$.

Expanding the differential with respect to x and setting $q + v = \mathrm{R}$, the rainfall rate, equation [14] is reduced to:

$$h[\theta + D\alpha]\partial C/\partial t = -qx\partial C/\partial x - RC \qquad [15]$$

An interesting feature of equation [15] is that a solution found for a certain scale of x can be simply converted to another scale of x by a scale factor. This is evident by substituting $X = xG$, where G is a scale factor, in equation [15]:

$$h[\theta + D\alpha]\partial C/\partial t = -qX\partial C/\partial X - RC \qquad [16]$$

which does not change the form of the equation. This means that if a solution is found for a slope length scale in centimeters, the solution can be transformed to a slope length scale in meters. Furthermore, the C in equation [16] can be replaced by relative concentration C/C_0, where C_0 is the initial concentration of the chemical in the source area. The results are, then, more generally applicable.

Equations [14] to [16] assume that all the overland flow from an upslope area passes over the downslope area. Actually, only a part of the upslope flow may pass over the downslope area, and the remainder shortcircuits to the outlet through a rill network. This phenomenon is approximated by multiplying the term $qX\partial C/\partial X$ in equation [16] by a factor F (<1.0), which represents the fraction of overland flow that passes along the downslope area at my given point. With this modification, equation [16] is rewritten as:

$$h[\theta + D\alpha]\partial C/\partial t = -FqX\partial C/\partial X - RC \quad [17]$$

This equation was solved numerically by a finite difference technique, subject to known initial conditions of chemical application on either the complete area, upper half area, or the lower half area of a field. The X and t increments were 0.25 and 0.125, respectively. With these small steps, the slope $\partial C/\partial X$ was estimated at the forward value of the time interval at each X grid point. Further reduction of the increments did not change the solution. The upper end of the field was taken as $X = 0$, with X increasing downslope. For entire-area and upper half-area application of a chemical, equation [17] at $X = 0$ is reduced to:

$$h[\theta + D\alpha]\partial C/\partial t = -RC \quad [18]$$

which can be integrated as

$$C = C_0 \exp[-Rt/h(\theta + D\alpha)] \quad [19]$$

where C_0 is the initial concentration of chemical in soil water within depth h. Equation [19] was used as the upper boundary condition for a sequential solving of equation [17] for points downslope. For the lower half-area application of the chemical, the boundary condition at point $X/2$ was analogous to equation [19]:

$$C = C_0 \exp[-(R + qX/2)t/h(\theta + D\alpha)] \quad [20]$$

The theoretical calculations for chemical application on the upper half area of sloping field showed that the chemical concentration in runoff is very small initially, increases with time, and later becomes equal to the concentration from a complete-area application. However, the partial-area application greatly reduces the relative amount of chemical lost to runoff. This reduction is greatest if the infiltration rate of the entire land area is high, but it is also appreciable if the infiltration rate of the lower area is increased. Ponding of water in the lower land area changes the shape of concentration-time curve, but it has a negligible effect on the total amount of chemical lost to runoff. Reduction of the effective soil depth of mixing in the lower land area greatly reduces the amount of an adsorbed chemical in runoff ($\alpha = 3$, Figure 17).

Shortcircuiting of chemical-laden runoff from the upper area to the outlet increases the amount of chemical in runoff. When the chemical is applied on the lower half area of the field, the shortcircuiting of chemical-free runoff from the upper half to the outlet reduces the amount of chemical lost (Figure 18).

The simple model described by equations [14] to [20], and its useful

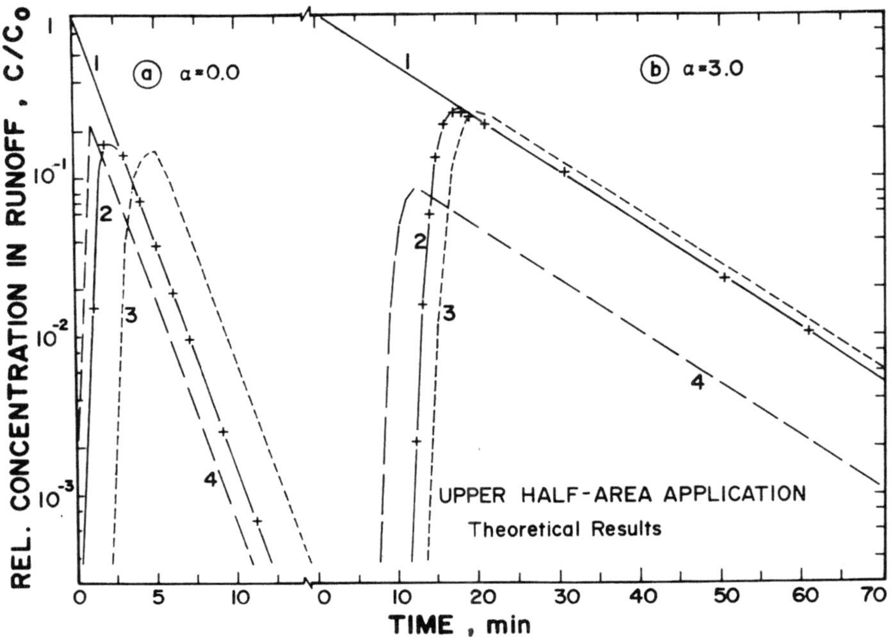

Figure 17. Relative concentration in runoff with time for a nonadsorbed chemical (a), and an adsorbed chemical (b) for complete and partial land-area applications. Curve 1: complete-area application with no ponding; curve 2: upper half-area application with no ponding; curve 3: upper half-area application with 2-mm depth of ponding over the lower half area; curve 4: same as curve 2 except the effective soil depth of interaction is reduced from 3 mm to 1 mm over the lower half area. The integrals of curves 2 and 3 on linear scale are about the same.

predictions given above, need to be evaluated under field conditions. A more detailed model, which incorporates the nonsteady overland flow in conjunction with the chemical mixing along the slope, has recently been developed by Woolhiser, Emmerich, and Hubbard (D. A. Woolhiser, personal communication, 1985). This model should also be evaluated.

VII. Shallow Subsurface Flow of Chemicals

Water that infiltrates into the soil, moves laterally above an impeding horizon, and then reappears on the soil surface or joins the surface runoff downslope has a high potential for transporting soil chemicals in solution into runoff. This type of flow is commonly known as interflow or subsurface runoff. This shallow subsurface flow may be significant in

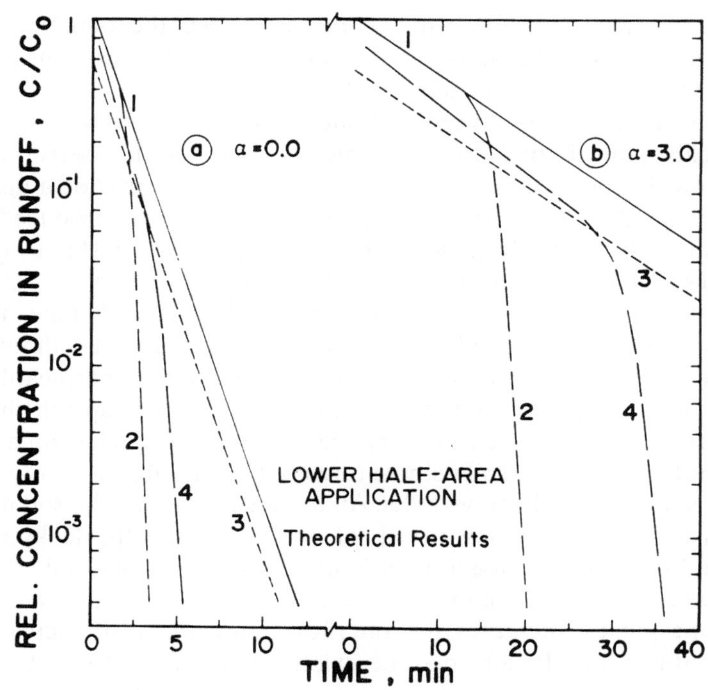

Figure 18. Calculated results for lower half-area application of chemicals. Curve 1: complete-area application; curve 2: lower half-area application; curve 3: same as curve 2 except all overland flow from upper half area is shortcircuited to the outlet; curve 4: same as curve 2 except half of the overland flow at each point on the lower half is shortcircuited. The integrals of curves 3 and 4 on linear scale are less than that of curve 2.

sloping soils with surface horizons of high permeability underlain by a horizon of much lower permeability (Hewlett and Hilbert, 1963; Whipkey, 1965; Jamison and Peters, 1967; Weyman, 1973; Pilgrim *et al.*, 1978). Root channels and other macropores in the surface soil could contribute substantially to storm flow (Mosley, 1979). The effect of interflow on the transfer of soil chemicals to surface runoff is well illustrated in a report by Barnett *et al.* (1972). In a bounded runoff plot on an agricultural soil subject to artificial rainfall, where all runoff was from interflow, average concentrations of N, K, Ca, Mg, S, and Cl were an order of magnitude greater as compared to similar plots on two other soils, where the runoff was mainly from overland flow. Burwell *et al.* (1976) and Owens *et al.* (1983) reported that shallow subsurface flow was the main pathway for soluble N transported to runoff from claypan soils of the midwestern United States. Hubbard and Sheridan (1983) reported similar results from Plinthic Coastal Plain soils, where 79% of the total runoff was subsurface flow, which contained 99% of the total N in runoff. Wu *et al.* (1983) found substantial amounts of pesticides transferred to runoff through shallow subsurface flow.

Delineation of the pathways and rates of interflow, and source areas of chemicals contributing to this flow, is essential to devise control measures. Theoretical analyses of the flow situations provide a means to accomplish this. Such analyses in layered soils have produced some interesting findings (Klute *et al.*, 1965; Ahuja *et al.*, 1981b; Ahuja and Ross, 1982, 1983). When the subsoil is completely impervious, the flow pathways in the topsoil are essentially parallel to the soil slope, and the chemical in the entire length of slope is subject to transport with interflow (Figure 19a). However, when the subsoil is even slightly permeable, with hydraulic conductivity as low as 1/100th that of the topsoil, the flow pathways and source areas are changed considerably. The slope length contributing to and the soil volume traversed by interflow are only a short distance upslope of the flow outlet. All the water entering the soil upslope of this distance flows downward (at an angle) into the leaky subsoil (Figure 19b). A leaky impeding horizon is more often a natural condition (Carlton *et al.*, 1985). The presence of a seepage face at the lower end of the slope greatly increases the interflow rates and also increases the contributing slope length to some extent (Ahuja and Ross, 1982). The actual contributing area and rate depend on relative hydraulic conductivities of topsoil and subsoil, depth of the topsoil, slope of the seepage face, and soil slope. In another situation, the presence of a layer of soil with a higher hydraulic conductivity than that of the topsoil between the topsoil and the impeding horizon increased severalfold the rate of interflow and the soil volume that was traversed by interflow (Ahuja and Ross, 1983). A recent field-plot study confirmed the theoretically predicted effect of leakage rates of the impeding horizon on the rates and plot surface area contributing to interflow (Lehman and Ahuja, 1985b). In a 70-cm-thick topsoil having a saturated hydraulic conductivity of 2.5 cm hr^{-1}, water and Br in interflow came from an average distance of 1.3 m upslope from the outlet face.

In many situations, the hydraulic conductivity of the top few centimeters of a topsoil may be extremely high compared with that of the layer below, which is fairly permeable. This could happen, for example, in a shallow tilled plot, cloddy plot, or a mulch-covered plot. Jamison and Peters (1967) measured hydraulic conductivity of nearly 850 cm hr^{-1} in the 0- to 2.5-cm depth interval of a soil and 49 cm hr^{-1} in the 2.5- to 5.0-cm interval. Appreciable interflow of water and chemicals can occur in this thin surface zone (Barnett *et al.*, 1972; Zaslavsky and Sinai, 1981).

Ridges and furrows on the soil surface and surface roughness result in interflow on a local scale. Bromide concentrations in runoff from small (1.7 cm high) ridges were nearly four times higher than those from a similarly treated unridged soil (Ahuja *et al.*, 1983).

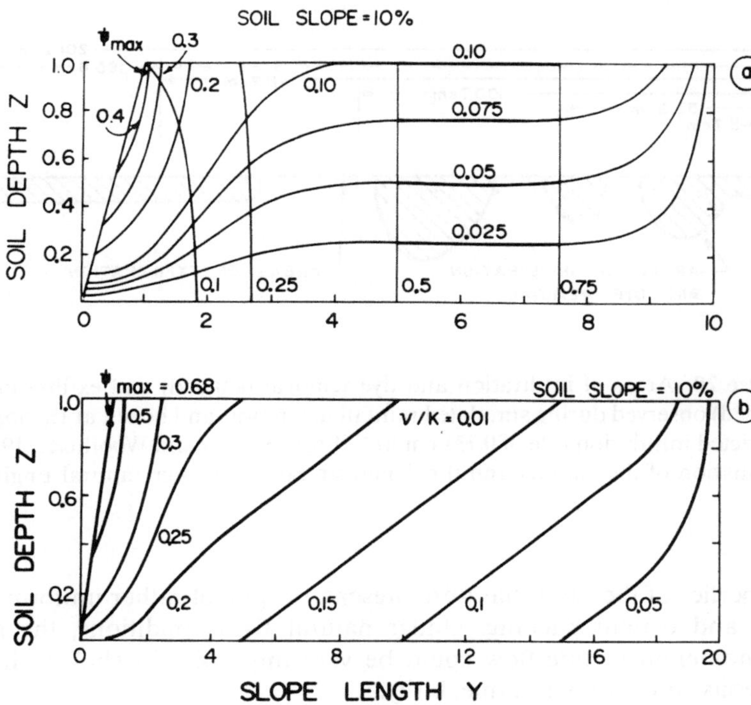

Figure 19.(a) Equipotential (vertical lines) and flow lines for interflow in a sloping soil with seepage face resting on an impermeable base. Slope length to soil depth ratio is 10 and slope of the seepage face 100%. *Y* and *Z* are dimensionless. Numbers on the flow lines are values of the dimensionless stream function, ψ/LK, with ψ_{max} the maximum value. The entire soil contributes to the interflow. Reproduced from Ahuja and Ross, *Soil Sci. Soc. Am. J.* 46:245–250, 1982, by permission of the Soil Science Society of America. **(b)** Flow lines in a sloping soil resting on a leaky base. The face slope is 100% and the leakage rate 1/100th the hydraulic conductivity of soil. Only the area left of the flow line = 0.2 contributes to interflow. Reproduced from Ahuja and Ross, *Soil Sci. Soc. Am. J.* 46:245–250, 1982, by permission of the Soil Science Society of America.

In heterogeneous shallow soils, the soil water may emerge and reenter at several places along the slope (Zaslavsky and Sinai, 1977; Steenhuis *et al.*, 1981). This phenomenon was clearly observed in the recent experiments of Snyder and Woolhiser (1985). In sand boxes saturated with a dye solution in the top layer, several areas of infiltration and exfiltration were seen along the 2-m length during the rainfall (Figure 20). These results were apparently caused by small nonhomogeneities in the hydraulic

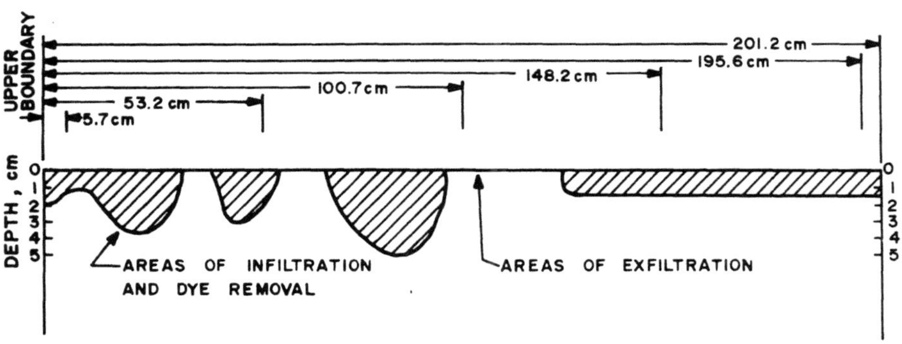

Figure 20. Areas of infiltration and dye removal (shaded) and exfiltration (not shaded) observed during simulated rainfall in Ottawa sand boxes, at 1% slope and restricted infiltration rate of 0.731 cm hr^{-1}. From Snyder and Woolhiser (1985), by permission of the authors and the American Society of Agricultural Engineers.

properties of the sand that were present, in spite of rather uniform grain size and careful packing. Under natural field conditions, the above phenomenon of interflow could be very important in shallow heterogeneous soils of high permeability.

VIII. Summary, Applications, and Suggested Further Work

In the past two decades, a considerable amount of useful information has been obtained from field experiments on the overall losses and concentrations of chemicals in runoff from different land uses, cropping systems, and management practices. The results show that weakly and moderately adsorbed pesticides and nutrients are transported primarily in solution form, and that this component of runoff may increase under some conservation tillage methods. A basic understanding of the transport process has begun to emerge more recently. Such understanding is needed for interpretation of observations and modeling the process, so as to devise measures for minimizing the losses. The models developed in the past are not quite satisfactory. This has been primarily due to our inability to accurately define the depth and degree of rainfall–soil interaction, adsorption–desorption relationships for chemicals, and subsurface pathways of water flow.

It has generally been assumed that a thin zone of surface soil mixes completely with rainwater during the transfer of a chemical from soil to runoff. Recent research indicates that a chemical may be transferred from the soil to runoff from a depth as great as 2.0 cm. However, both the degree of mixing between soil and rainwater and the chemical transfer decrease exponentially with depth. An effective depth of complete and uniform mixing may be approximately assumed under high infiltration conditions, but not where low infiltration rates occur. Chemical transferred to runoff is much higher with restricted infiltration. The depth and degree of mixing, and hence the chemical transfer, are increased by an increase in rainfall kinetic energy, intensity, slope, and permeability of the soil surface layer.

Laboratory batch studies indicate that the desorption of moderately adsorbed pesticides can approach equilibrium in minutes, whereas for P the release is slow and more time dependent. A power-form time dependence of P desorption is shown to be widely applicable. However, the desorption characteristics of soil need to be determined under rainfall conditions, taking into the account the variation in the degree of mixing with depth, the processes involving physical absorption and retention by soil, and the effects of a surface residue or plant cover.

A nonuniform mixing model, which incorporates the variable degree of mixing with depth, the effect of infiltration rate, and a simple proportional relation for desorption, is shown to describe adequately experimental data for adsorbed, moderately adsorbed, and mobile chemicals. An empirical power-form model also appears promising. A simple theory is presented for analysis of partial land-area applications of chemicals and related to the effects of buffer strips.

Shallow subsurface flow or interflow can be an important pathway for the transfer of soluble chemicals in runoff from claypan and plinthic soils, and even in normal soils when permeability of the surface zone is very high. However, if the impeding layer is not completely impervious, the contributing area for chemicals is restricted to a small distance upslope of the watershed outlet. A layer of high permeability above the impeding layer greatly increases the contributing area and flow rates. In highly heterogeneous soils and in soils with rough surface and ridges, shallow interflow can occur at several places along the slope.

Based on the research done so far, some possible ways to reduce chemical loss in runoff are to: (1) incorporate chemicals into the soil or apply them below the surface rather than on the surface; (2) use sprinkler irrigation, when feasible, to water in the surface-applied chemicals; (3) minimize the area of application by specific placement of chemicals; (4) maintain and enhance infiltration rate of the soil; (5) increase permeability of the claypan layer by subsoiling and growing deep-rooted crops; (6) use buffer strips at the lower end of a slope, maintaining good surface cover and high infiltration; and (7) intercept runoff from an upslope area where no chemical has been applied and divert it away from an area where a chemical has been applied.

Further research is needed on the following basic aspects of the chemical transfer processes discussed in this contribution: (1) effects of unsteady overland flow and slope length; (2) effect of crop canopy and residue cover; (3) partial land-area application of chemicals and desirable characteristics of a buffer strip; (4) kinetics of chemical release under actual rainfall and varying soil conditions; (5) interflow in partially saturated shallow and heterogeneous soils; and (6) effects of tillage, surface shaping, crusting-sealing, and residue management practices.

Acknowledgments

I am very grateful to my colleagues, Andrew Sharpley, Oliver Lehman, and Gary Heathman, with whom I have worked in the area of chemical transfer to runoff during the past few years and who have contributed to some of the interpretations and ideas presented in this paper. I have used our joint work profusely for illustrations. I am also grateful to J. K. Snyder and D. A. Woolhiser for permitting me to reproduce graphs from their paper, and to J. L. Baker for providing his experimental data used in the analyses.

References

Ahuja, L. R. 1982. Release of a soluble chemical from soil to runoff. *Trans. ASAE* 25:948–956, 960.

Ahuja, L. R., and O. R. Lehman. 1983. The extent and nature of rainfall-soil interaction in the release of soluble chemicals to runoff. *J. Environ. Qual.* 12:34–40.

Ahuja, L. R., and J. D. Ross. 1982. Interflow of water through a sloping soil with seepage face. *Soil Sci. Soc. Am. J.* 46:245–250.

Ahuja, L. R., and J. D. Ross. 1983. Effect of subsoil conductivity and thickness on interflow pathways, rates and source areas for chemicals in a sloping layered soil with seepage face. *J. Hydrol.* 64:189–204.

Ahuja, L. R., A. N. Sharpley, M. Yamamoto, and R. G. Menzel. 1981a. The depth of rainfall–runoff–soil interactions as determined by ^{32}P. *Water Resour. Res.* 17:969–974.

Ahuja, L. R., J. D. Ross, and O. R. Lehman. 1981b. A theoretical analysis of interflow of water through surface soil horizons with implications for movement of chemicals in field runoff. *Water Resour. Res.* 17:65–72.

Ahuja, L. R., A. N. Sharpley, and O. R. Lehman. 1982a. Effect of soil slope and rainfall characteristics on phosphorus in runoff. *J. Environ. Qual.* 11:9–13.

Ahuja, L. R., A. N. Sharpley, R. G. Menzel, and S. J. Smith. 1982b. Modeling the release of phosphorus and related adsorbed chemicals from soil to overland flow. In Singh, V. P. (Ed.), *Modeling Components of Hydrologic Cycle*. Water Resources Publications, Fort Collins, CO, pp. 463–484.

Ahuja, L. R., O. R. Lehman, and A. N. Sharpley. 1983. Bromide and phosphate in runoff water from shaped and cloddy soil surfaces. *Soil Sci. Soc. Am. J.* 47:746–748.

Ahuja, L. R., G. C. Heathman, and O. R. Lehman. 1985. Transport of Chemicals in Runoff from Partial Land-Area Applications. Tech. Rept., USDA-ARS, Water Quality and Watershed Research Laboratory, Durant, OK, USA.

Amer, F., D. R. Boulden, C. A. Black, and F. R. Duke. 1955. Characterization of soil phosphorus by anion exchange resin and ^{32}P equilibration. *Plant Soil* 6:391–408.

Asmussen, L. E., A. W. White, E. W. Hauser, and J. A. Sheridan. 1977. Reduction of 2,4-D load in surface runoff down a grassed waterway. *J. Environ. Qual.* 6:159–162.

Bailey, G. W., R. R. Swank, Jr., and H. P. Nicholson. 1974. Predicting pesticide runoff from agricultural lands: A conceptual model. *J. Environ. Qual.* 3:95–102.

Baker, J. L., and H. P. Johnson. 1979. The effect of tillage systems on pesticides in runoff from small watersheds. *Trans. ASAE* 22:554–559.

Baker, J. L., and J. M. Laflen. 1979. Runoff losses of surface-applied herbicides as affected by wheel tracks and incorporation. *J. Environ. Qual.* 8:602–607.

Baker, J. L., and J. M. Laflen. 1982. Effects of corn residue and fertilizer management on soluble nutrient runoff losses. *Trans. ASAE* 25:344–348.

Baker, J. L., J. M. Laflen, and H. P. Johnson. 1978. Effect of tillage systems on runoff losses of pesticides. A rain simulation study. *Trans. ASAE* 21:886–892.

Baker, J. L., J. M. Laflen, and R. O. Hartwig. 1982. Effects of corn residue and herbicide placement on herbicide runoff losses. *Trans. ASAE* 25:340–343.

Barisas, S. G., J. L. Baker, H. P. Johnson, and J. M. Laflen. 1978. Effect of tillage systems on runoff losses of nutrients. *Trans. ASAE* 21:893–897.

Barnett, A. P., J. R. Carreker, F. Abruna, W. A. Jackson, A. E. Dooley, and J. H. Holladay. 1972. Soil and nutrient losses in runoff with selected cropping treatments on tropical soils. *Agron. J.* 64:391–395.

Barrow, N. J. 1979. The description of desorption of phosphate from soil. *J. Soil Sci.* 30:259–270.

Baur, J. R., R. W. Bovey, and M. G. Merkle. 1972. Concentration of picloram in runoff water. *Weed Sci.* 20:309–313.

Bruce, R. R., L. A. Harper, R. A. Leonard, W. M. Synder, and A. W. Thomas. 1975. A model for runoff of pesticides from small upland watersheds. *J. Environ. Qual.* 4:541–548.

Burwell, R. E., G. E. Schuman, K. E. Saxton, and H. G. Heinemann. 1976. Nitrogen in subsurface drainage from agricultural watersheds. *J. Environ. Qual.* 5:325–329.

Carlton, W. L., H. F. Perkins, and R. A. Leonard, 1985. Movement of water in a plinthic Paleudult using a bromide tracer. *Soil Sci.* 139:62–66.

Chien, S. H., and W. R. Clayton. 1980. Application of Elovich equation to the kinetics of phosphate release and sorption of soils. *Soil Sci. Soc. Am. J.* 44:265–268.

Cooke, I. J. 1966. A kinetic approach to the description of soil phosphate status. *J. Soil Sci.* 17:56–64.

Crank, J. 1964. *The Mathematics of Diffusion*. Oxford University Press, London, 347 pp.

Crawford, N. H., and A. S. Donigian, Jr. 1973. Pesticide Transport and Runoff Model for Agricultural Lands. EPA 660/2-73-013. Southeast Environmental Research Laboratory, U.S. Environmental Protection Agency, Athens, GA, 211 pp.

Davis, H. H., Jr., and A. S. Donigian, Jr. 1979. Simulating nutrient movement and transformations with the ARM model. *Trans. ASAE* 22:1081–1086.

Donigian, A. S., Jr., and N. H. Crawford, 1976. Modeling Pesticides and Nutrients on Agricultural Lands. EPA 600/2-76-043. Environmental Research Laboratory, U.S. Environmental Protection Agency, Athens, GA, 317 pp.

Donigian, A. S., Jr., D. C. Beyerlein, H. H. Davis, and N. H. Crawford. 1977. Agricultural Runoff Management (ARM) Model Version II: Refinement and Testing. EPA 600/3-77-098. Environmental Research Laboratory, U.S. Environmental Protection Agency, Athens, GA, 293 pp.

Drynan, W. R. 1980. Sources of pollution to the Great Lakes. In: *Seminar on Water Quality Management Tradeoffs*. EPA 905/9-80-009. U.S. Environmental Protection Agency, Chicago, IL, pp. 63–73.

Duda, A. M. 1982. Municipal point source and agricultural nonpoint source contributions to coastal eutrophication. *Water Resour. Bull.* 18:397–407.

Duda, A. M., and D. S. Finn. 1983. Influence of livestock on nonpoint source nutrient levels of streams. *Trans. ASAE* 26:1710–1716.

Edwards, W. M., G. B. Triplett, Jr., and R. M. Kramer. 1980. A watershed study of glyphosate transport in runoff. *J. Environ. Qual.* 9:661–665.

Enfield, C. G., and B. E. Bledsoe. 1975. Kinetic Model for Orthophosphate Reactions in Mineral Soils. EPA Technology Series, Rep. No. 660/2-75-002. Environmental Research Laboratory, U.S. Environmental Protection Agency, Ada, OK.

Evans, R. L., and J. J. Jurinak. 1976. Kinetics of phosphate release from a desert soil. *Soil Sci.* 121:205–211.

Foster, G. R., L. D. Meyer, and C. A. Onstad. 1977. An erosion equation derived from basic erosion principles. *Trans. ASAE* 20:678–682.

Frere, M. H., C. A. Onstad, and H. N. Holtan. 1975. ACTMO, an Agricultural Chemical Transport Model. ARS-H-3, U.S. Dept. Agr., Hyattsville, MD, 56 pp.

Frere, M. H., J. D. Ross, and L. J. Lane. 1980. The nutrient submodel. Chapter 4. In: Knisel, W. G. (ed.), *CREAMS: A Field Scale Model For Chemicals, Runoff, and Erosion from Agricultural Management Systems*. U.S. Dept. Agr. Conservation Res. Rept. No. 26, pp. 65–87, Washington, D.C.

Haith, D. A. 1980. A mathematical model for estimating pesticide losses in runoff. *J. Environ. Qual.* 9:428–433.

Haith, D. A., and L. J. Tubbs. 1981. Watershed loading functions for nonpoint sources. *J. Environ. Eng. Div. ASCE* 107:121–137.

Hall, J. K., N. L. Hartwig, and L. D. Hoffman. 1983. Application mode and alternate cropping effects on atrazine losses from a hillside. *J. Environ. Qual.* 12:336–340.

Haque, R., and R. Sexton. 1968. Kinetic and equilibrium study of the adsorption of 2,4-Dichlorophenoxy acetic acid on some surfaces. *J. Colloid Inter. Sci.* 27:818–827.

Haque, R., F. T. Lindstrom, V. H. Freed, and R. Sexton. 1968. Kinetic study of sorption of 2,4-D on some clays. *J. Environ. Sci. Technol.* 2:207–211.

Heathman, G. C., L. R. Ahuja, and O. R. Lehman. 1985a. The transfer of soil surface-applied chemicals to runoff. *Trans. ASAE* 28:1909–1915, 1920.

Heathman, G. C., L. R. Ahuja, and J. L. Baker. 1985b. A Non-uniform Mixing Model for Pesticide Transfer to Runoff Water. Tech. Rept., USDA-ARS, Water Quality and Watershed Research Laboratory, Durant, OK, USA. To appear in *Transactions of the ASAE*.

Hewlett, J. D., and A. R. Hibbert. 1963. Moisture and energy conditions within a sloping mass during drainage. *J. Geophys. Res.* 68:1081–1087.

Huff, D. D., and P. Kruger. 1967. The chemical and physical parameters in a hydrologic transport model for radioactive aerosols. *Proc. Int. Hydrol. Symp.* 1:128–135.

Hubbard, R. K., and J. M. Sheridan. 1983. Water and nitrate-nitrogen losses from a small, upland, Coastal Plain Watershed. *J. Environ. Qual.* 12:291–295.

Hubbard, R. K., A. E. Erickson, B. G. Ellis, and A. R. Wolcott. 1982. Movement of diffuse source pollutants in small agricultural watersheds of the Great Lakes Basin. *J. Environ. Qual.* 11:117–123.

Ingram, F. M. 1976. Disposition of bromide applied to grass and fallow plots. *Proc. Okla. Acad. Sci.* 56:38–41.

Ingram, J. J., and D. A. Woolhiser. 1980. Chemical transfer into overland flow. In: Proc. Symp. Watershed Management, Boise, ID, July 21-23. Am. Soc. Civ. Eng., New York, pp. 40–53.

Jamison, V. C., and D. B. Peters. 1967. Slope length of claypan soil affects runoff. *Water Resour. Res.* 3:471–480.

Johnson, H. P., J. L. Baker, W. D. Schrader, and J. M. Laflen. 1979. Tillage system effects on sediment and nutrients in runoff from small watersheds. *Trans. ASAE* 22:1110–1114.

Kissel, D. E., C. W. Richardson, and E. Burnett. 1976. Losses of nitrogen in surface runoff in the Blackland Prairie of Texas. *J. Environ. Qual.* 5:288–293.

Klute, A., E. J. Scott, and F. D. Whisler. 1965. Steady state water flow in a saturated inclined slab. *Water Resour. Res.* 1:287–294.

Knisel, W. G. (Ed.). 1980. CREAMS: A Field Scale Model for Chemicals, Runoff, and Erosion from Agricultural Management Systems. U.S. Dept. Agr. Conservation Res. Rept. 26, 640 pp.

Kuo, S., and E. G. Lotse. 1974. Kinetics of phosphate adsorption and desorption by hematite and gibbsite. *Soil Sci.* 116:400–406.

Laflen, J. M., and M. A. Tabatabai. 1984. Nitrogen and phosphorus losses from corn-soybean rotations as affected by tillage practices. *Trans. ASAE* 27:58–63.

Leenheer, J. A., and J. L. Ahlrichs. 1971. A kinetic and equilibrium study of the adsorption of carbaryl and parathion upon soil organic matter. *Soil Sci. Soc. Am. Proc.* 35:700–705.

Lehman, O. R., and L. R. Ahuja. 1985a. Effects of Slope Length and Rainfall Intensity on the Transfer of Nonadsorbed and Adsorbed Chemicals to Runoff. Tech. Rept., USDA-ARS, Water Quality and Watershed Research Laboratory, Durant, OK, USA.

Lehman, O. R., and L. R. Ahuja. 1985b. Interflow of water and tracer chemical on sloping field plots with exposed seepage faces. *J. Hydrol.* 76:307–317.

Leonard, R. A., and R. D. Wauchope. 1980. The pesticide submodel. Chapter 5. In: Knisel, W. G. (ed.), *CREAMS: A Field Scale Model for Chemicals, Runoff and Erosion from Agricultural Management Systems.* U.S. Dept. Agr. Conservation Res. Rept. No. 26, pp. 88–112, Washington, D.C.

Leonard, R. A., G. W. Langdale, and W. G. Fleming. 1979. Herbicide runoff from upland piedmont watersheds—data and implications for modeling pesticide transport. *J. Environ. Qual.* 8:223–229.

Lindstrom, F. T., R. Haque, and W. R. Coshow. 1970. Adsorption from solution.

III. A new model for the kinetics of adsorption-desorption processes. *J. Phys. Chem.* 74:495–502.

McDowell, L. L., and K. C. McGregor. 1984. Plant nutrient losses in runoff from runoff from no-till soyabeans. *Trans. ASAE.*23:643–648.

McDowell, L. L., and K. C. McGregor. 1984. Plant nutrient losses in runoff from conservation tillage corn. Soil Tillage Res 4:79–91.

McDowell, L. L., J. D. Schreiber, and H. B. Pionke. 1980. Estimating soluble (PO_4-P) and labile phosphorus in runoff from croplands. Chapter 14. In: Knisel, W. G. (ed.), *CREAMS: A Field Scale Model for Chemicals, Runoff, and Erosion from Agricultural Management Systems.* U.S. Dept. Agr. Conservation Res. Rept. No. 26, pp. 509–533, Washington, D.C.

McDowell, L. L., G. H. Willis, C. E. Murphree, L. M. Southwick, and S. Smith. 1981. Toxaphene and sediment yields in runoff from a Mississippi Delta Watershed. *J. Environ. Qual.* 10:120–125.

McDowell, L. L., G. H. Willis, and C. E. Murphree. 1984. Plant nutrient yields in runoff from a Mississippi Delta Watershed. *Trans. ASAE* 27:1059–1066, 1073.

Menzel, R. G., E. D. Rhoades, A. E. Olness, and S. J. Smith. 1978. Variability of annual nutrient and sediment discharges in runoff from Oklahoma cropland and rangeland. *J. Environ. Qual.* 7:401–406.

Miller, B. A., and T. C. Daniel. 1981. An empirical equation for predicting soluble phosphorus loads from agricultural watersheds. *Trans. ASAE* 24:375–381, 390.

Miller, M. H., J. B. Robinson, D. R. Coate, A. C. Spires, and D. W. Draper. 1982. Agriculture and water quality in the Canadian Great Lakes Basin. III. Phosphorus. *J. Environ. Qual.* 11:487–493.

Monke, E. J., D. W. Nelson, D. B. Beasley, and A. B. Bottcher. 1981. Sediment and nutrient movement from the Black Creek watershed. *Trans. ASAE* 24:391–395, 400.

Mosley, P. 1979. Streamflow generation in a forested watershed, New Zealand. *Water Resour. Res.* 15:795–806.

Nielsen, D. R., and J. W. Biggar. 1963. Miscible displacement. III. Theoretical considerations. *Soil Sci. Soc. Am. Proc.* 26:216–221.

Olness, A., S. J. Smith, E. D. Rhoades, and R. G. Menzel. 1975. Nutrient and sediment discharge from agricultural watersheds in Oklahoma. *J. Environ. Qual.* 4:331–336.

Owens, L. B., R. W. van Keuren, and W. M. Edwards. 1983. Nitrogen loss from a high-fertility, rotational pasture program. *J. Environ. Qual.* 12:346–350.

Pilgrim, D. H., D. D. Huff, and T. D. Steele. 1978. A field evaluation of subsurface and surface runoff: II. Runoff processes. *J. Hydrol.* 38:319–341.

Ponce, S. L. 1975. Examination of a Nonpoint Source Loading Function for Marcos Shale Wildlands of the Prince River Basin, Utah. Ph.D. Dissertation, Utah State University, Logan.

Posner, A. M., and J. W. Bowden. 1980. Adsorption isotherms: Should they split? *J. Soil Sci.* 31:1–10.

Romkens, M. J. M., and D. W. Nelson. 1974. Phosphorus relationships in runoff from fertilized soils. *J. Environ. Qual.* 3:10–13.

Romkens, M. J. M., D. W. Nelson, and J. V. Mannering. 1973. Nitrogen and phosphorus composition of surface runoff as affected by tillage method. *J. Environ. Qual.* 2:292–295.

Scheppers, J. S., B. L. Hackes, and D. D. Francis. 1982. Chemical water quality of runoff from grazing land in Nebraska: II. Contributing factors. *J. Environ. Qual.* 11:355–359.

Schreiber, J. D. 1985. Leaching of nitrogen, phosphorus, and organic carbon from wheat straw residues: II. Loading rate. *J. Environ. Qual.* 14:256–260.

Schreiber, J. D. and L. L. McDowell. 1985. Leaching of nitrogen, phosphorus, and organic carbon from wheat straw residues: I. Rainfall intensity. *J. Environ. Qual.* 14:251–256.

Schuman, G. E., R. G. Spomer, and R. F. Piest. 1973a. Phosphorus losses from four agricultural watersheds in Missouri Valley loses. *Soil Sci. Soc. Am. Proc.* 37:424–427.

Schuman, G. E., R. E. Burwell, R. F. Piest, and R. G. Spomer. 1973b. Nitrogen losses in surface runoff from agricultural watersheds in Missouri Valley loess. *J. Environ. Qual.* 2:299–302.

Sharpley, A. N. 1981. The contribution of phosphorus leached from crop canopy to losses in surface runoff. *J. Environ. Qual.* 10:160–165.

Sharpley, A. N., J. K. Syers, and R. W. Tillman. 1978. An improved soil sampling procedure for the prediction of dissolved inorganic phosphate concentrations in surface runoff from pasture. *J. Environ. Qual.* 7:455–456.

Sharpley, A. N., L. R. Ahuja, M. Yamamoto, and R. G. Menzel. 1981a. The kinetics of phosphorus desorption from soil. *Soil Sci. Soc. Am. J.* 45:493–496.

Sharpley, A. N., L. R. Ahuja, and R. G. Menzel. 1981b. The release of soil phosphorus to runoff in relation to the kinetics of desorption. *J. Environ. Qual.* 10:386–391.

Sharpley, A. N., S. J. Smith, and R. G. Menzel. 1982. Prediction of phosphorus losses in runoff from Southern Plains watersheds. *J. Environ. Qual.* 11:247–251.

Siemens, J. C., and W. R. Oschwald. 1978. Corn-soybean tillage systems: Erosion control, effects on crop production, costs. *Trans. ASAE* 21:293–302.

Snyder, J. K., and D. A. Woolhiser. 1985. Effects of infiltration on chemical transport into overland flow. *Trans. ASAE* (in press).

Steenhuis, T. S., and M. F. Walter. 1980. Closed form solution for pesticide loss in runoff water. *Trans. ASAE* 23:615–620, 628.

Steenhuis, T. S., R. E. Muck, and T. R. McCarty. 1981. Conceptualization of Water, Nutrient, and Pesticides Transport in Shallow Soils. Tech. Rept., Dept. Agr. Eng., Cornell Univ., Ithaca, NY, 17 pp.

Stewart, B. A., D. A. Woolhiser, W. H. Wischmeier, J. H. Caro, and M. H. Frere. 1975. Control of Water Pollution from Cropland. Vol. I. A Manual for Guideline Development. U.S. Dept. Agr. Rept. ARS-H-5-1, 111 pp.

Stewart, B. A., D. A. Woolhiser, W. H. Wischmeier, J. H. Caro, and M. H. Frere. 1976. Control of Water Pollution from Cropland. Vol. II. An Overview. U.S. Dept. Agr. Rept. ARS-H-5-2, 187 pp.

Travis, C. C., and E. L. Etnier. 1981. A survey of sorption relationships for reactive solutes in soil. *J. Environ. Qual.* 10:8–17.

Trichell, D. W., H. L. Morton, and M. G. Merkle. 1968. Loss of herbicides in runoff water. *Weed Sci.* 16:447–449.

U.S. Environmental Protection Agency (EPA). 1980. Lake Restoration in Cobbossee Watershed. EPA 625/2-80-027. Cincinnati, OH, 15 pp.

Vig, A. C., and G. Dev. 1979. Kinetics of phosphate desorption in soils. *J. Nucl. Agr. Biol.* 8:62–66.

Wahid, P. A., and N. Sethunathan. 1978. A simple method to study pesticide sorption in soils at short time intervals. *Soil Sci.* 126:56–58.

Wauchope, R. D. 1978. The pesticide content of surface water draining from agricultural fields—a review. *J. Environ. Qual.* 7:459–472.

Wauchope, R. D., and R. A. Leonard. 1980. Maximum pesticide concentrations in agricultural runoff: A semiempirical prediction formula. *J. Environ. Qual.* 9:665–672.

Wauchope, R. D., and R. S. Myers. 1985. Adsorption-desorption kinetics of atrazine and linuron in freshwater-sediment acqueous slurries. *J. Environ. Qual.* 14:132–136.

Weyman, D. R. 1973. Measurement of the downslope flow of water in a soil. *J. Hydrol.* 20:267–288.

Whipkey, R. Z. 1965. Subsurface storm flow from forested slopes. *Bull. Int. Assoc. Sci. Hydrol.* 10:74–85.

Wiese, A. F., K. E. Savage, J. M. Chandlers, L. C. Liu, L. S. Jeffery, J. B. Weber, and K. S. Lafleur. 1980. Loss of fluometuron in runoff water. *J. Environ. Qual.* 9:1–5.

Williams, J. R., and R. W. Haan, Jr. 1978. Optimal Operation of Large Agricultural Watersheds with Water Quality Constraints. Tech. Rept. No. 96. Water Resour. Res. Inst., Texas A & M Univ., College Station, TX, 1520 pp.

Wu, T. L., D. L. Correll, and H. E. H. Remenapp. 1983. Herbicide runoff from experimental watersheds. *J. Environ. Qual.* 12:330–336.

Zaslavsky, D., and G. Sinai. 1977. Surface hydrology. Nota 1017. Instituut voor Cultuur Techniek and Waterhuishouding, Wageningen, The Netherlands.

Zaslavsky, D., and G. Sinai. 1981. Surface hydrology: III. Causes of lateral flow. *J. Hydraul. Div. ASCE* 107(HY1):37–52.

Agronomical and Ecological Impact of Irrigation on Soil and Water Salinity

I. Szabolcs*

I. Introduction

The agricultural and industrial activities of mankind have a substantial impact on the environment. Production systems—agriculture, forestry,

*Research Institute for Soil Science and Agricultural Chemistry of the Hungarian Academy of Sciences, Budapest, Hungary.

© 1986 by Springer-Verlag New York Inc.
Advances in Soil Science, Volume 4

fishery, etc., that parallel the production of food and raw materials—change the properties of soil, water, air, and other components of the biosphere with respect to alterations caused by production in the environment. Agriculture has a particular importance. The growing world population, development of techniques, and extension of agricultural land have resulted in increasing change in nature. During the history of mankind, the human effect has always been significant, but it has also had serious consequences in many places of the world. By now this effect has become a worldwide problem and many national and international organizations are confronted with the consequences.

Irrigation, which is one of the oldest methods of agriculture, also has a long history of different results, favorable and unfavorable. Among the latter the problem of salinity has been and remains one of the most important (USSR-UNEP) Project, 1982; Shainberg and Shalhevet, 1984).

Secondary salinization is the most harmful and extended phenomenon among the effects of irrigation on the soils and environment. This process (secondary because caused by human action) has been known since ancient times and its results are still with us today. More than half of all the irrigated soils of the world are affected by secondary salinization and/or alkalization (Figure 1).

Aubert (1962), Buringh (1960), Kovda (1947, 1980), Rozanov (1961), Szabolcs (1979), Worthington (1977), and many other authors have described the consequences of the expanding salinity in the irrigated areas of the world which has developed because of primitive or improper technology.

For many thousands of years irrigation often caused the accumulation of harmful salts, in both soils and waters, and has a detrimental effect, leading to the decline not only of agricultural production but also of ancient cultures and civilizations.

II. Salinity and Secondary Salinization in Agriculture in Ancient Times

Particularly in arid and semiarid regions, irrigation is as old as agriculture itself. The lack of rainfall made imperative the application of irrigation in many ancient agricultural systems. The problem stemmed from the fact that such systems developed mainly in arid regions where, according to the landscape geochemistry of deserts and semideserts, salts tend to accumulate. This problem occurred with nearly all of the ancient cultures employing irrigation (Dregne, 1976; Hodge and Duisberg, 1963, Kovda *et al.*, 1967).

During the whole history of the use of irrigation in dry countries, the relationship between irrigation and salinization has never been fully

Figure 1. Secondary alkalinization of irrigated soil.

elaborated in a comprehensive volume. However, many books and papers describing adverse effects and their consequences have been published (Hyams, 1976).

It is well known that in old Mesopotamia, an area of fertile soil between the Tigris and Euphrates river, quantities of grain and other crops sufficient to feed large populations were produced for a long time. In modern times, this area has been a bare desert (Buringh, 1960).

It is also well known that in ancient China, in the Indus Valley, and in South America vast territories have turned into deserts because of salinity problems resulting from irrigation by ancient societies. The problem of secondary salinization runs through the whole history of mankind. Evidently there was neither sufficient knowledge nor technical means to predict, explain, and combat salinization for many thousands of years. Consequently, the degradation of the fertility of these soils and other adverse effects were recognized too late to prevent them from ruin (Szabolcs, 1979). The process forced people to leave the land as it became saline, and to shift the irrigation to another place, which, in many cases, also became salinized. As long as new territories were available, the shifting of irrigated agriculture offered a temporary solution. However, as population density increased or new lands were exhausted, tragic consequences became apparent. More than one such example is known from history (FAO, 1976).

In order to estimate the global development and extension of secondary salinization and alkalization, as well as to predict the possible consequences of the increase of irrigation, the causes and dynamics of the process must be characterized.

III. The Main Aspects of Secondary Salinization and Alkalization

It is necessary to briefly summarize the main aspects of salinization and alkalinization caused by irrigation and the agricultural as well as environmental consequences.

There are only a few chemical elements and compounds that play a decisive role in the salinization and alkalization of rocks, soils, and waters. They are as follows:

Cations: Ca^{2+}, Na^{+}, Mg^{2+}, K^{+}

Anions: Cl^{-}, $[SO_4]^{2-}$, $[HCO_3]^{-}$, $[SiO_3]^{2-}$, $[CO_3]^{2-}$.

Here and in this contribution generally, the acid sulfate soils and the problem of some toxic elements occurring in the different saline and alkali soils are not discussed. Only the dominant elements and compounds regulating the process mentioned are interpreted.

Arany (1956), Bazilevich (1965), Fersman (1934), Kovda (1947), Polynov (1956), and Szabolcs (1974) have described the influence of geochemical mass flow on the soil-forming processes, particularly on salinity and alkalinity. Diverse processes of salt migration may develop because of different climatic, lithologic, and soil conditions in different areas, effecting the diversity of morphology, physics, and chemistry of saline and alkali soils of different areas.

In this review only the main rules and regularities of salt accumulation and their influence on soil-forming processes are discussed.

It is evident that in arid and semiarid areas the weathering processes result, in water-soluble compounds, as a final product, because of the lack of precipitation, are not removed from the place of their formation. As discussed earlier, these final products are mainly responsible for the salt accumulation in rocks, soils, and waters.

In Table 1, according to Fersman (1934), the places of the dominant elements in the sequences of extraction are demonstrated. The sequences with growing numbers indicate the decreasing mobility of elements during the weathering processes.

Table 1. Sequences of ion extraction during weathering

Sequence of extraction	Ions	Energy coefficient
I	Cl^-, Br^-	0.23
	$[NO_3]^-$	0.18
	$[SO_4]^{2-}$	0.66
	$[CO_3]^{2-}$	0.77
II	Na^+	0.45
	K^+	0.36
	Ca^{2+}	1.75
	Mg^{2+}	2.10
III	$[SiO_3]^{2-}$	2.75
IV	Fe^{3+}	5.15
	Al^{3+}	4.25

According to Fersman (1934), an enrgy coefficient can be calculated on the basis of known lattice energies in inorganic salts. These values are called "experimental energy coefficients." The energy coefficients of Fersman (1934) are closely related to the sequence of the extraction of ions from minerals, to the rate of migration of ions, and to their ability to accumulate in sediments and soils (Salinity Problems, 1961).

The weathering of rocks has been the primary means of soluble salts getting into natural waters, sediments, and soils (Borovsky and Pogrebinsky, 1964).

The geochemistry of salts in a certain place is determined by the mobility of the compounds formed and by the sequence of the precipitation of weathering products. The mobility of the rock-forming elements depends on the following factors:

the stability of the crystalline network
the radius of ions formed during weathering
the charge of the ions formed during weathering

It is evident that the possibilities of translocation for weathering products depend mainly on their mobility.

From Table 1 it follows that the elements and compounds with a dominant role in salinization and alkalization are mainly in sequences I and II; in other words, they are capable of intensive migration. In spite of this, very diverse values can be measured as to the mobility of the compounds mentioned, and their occurrences in rocks and waters are similarly diverse. In Table 2 the relative mobilities and ratios of nine elements and ions are set out according to Polynov (1956).

Table 2. Relative mobility and average distribution of some compounds and ions in rocks and waters

Compounds	Average content in igneous rocks (g/kg)	Average content in mineral residue of river waters (g/kg)	Relative mobility[a] (%)
SiO_2	590	128	0.20
Al_2O_3	153	9	0.02
Fe_2O_3	73	4	0.04
Ca^{2+}	36	147	3.00
Mg^{2+}	21	49	1.30
Na^+	30	95	2.40
K^+	26	44	1.25
Cl^-	0.5	67	100.00
$[SO_4]^{2-}$	1.5	116	57.00
$[CO_3]^{2-}$	—	365	—

[a]Relative to chloride

Accumulated salts may be found in many places, particularly in soils and waters of dry areas, and constitute a potential hazard for irrigated agriculture.

The other source of salts may be surface waters that are used for irrigation purposes. In this respect the picture is rather multicolored in arid and semiarid regions. While in many places, particularly in big rivers, the water is of good quality (i.e., it has a low salt concentration), in other places, particularly in small tributaries, lakes, lagoons, and swamps, saline water often exists. Sooner or later, generally, when saline water is utilized for irrigation, secondary salinization and/or alkalinization results. Here, it should be mentioned that on seashores and in related areas the salt content of sea water has a direct or indirect effect on the salinity of adjoining areas (Kovda and Minashina, 1967).

Remarkable differences can also be found in the chemical compositions of river waters, on the different continents (Table 3). River waters in South America and Australia are twice as dilute as the global average and nearly three times as dilute as the waters of Europe. Even larger differences can be observed in the concentration of some cations and anions in river waters. However, it is also clear from this table that the river waters constitute good-quality irrigation water on all continents, at least insofar as their average values are concerned (Szabolcs and Darab, 1955).

On the other hand, sharp differences exist among the chemical compositions of the waters of different rivers. In Table 4 the salinity levels of the Volga and Amu Daria rivers are compared (Kovda, 1947). While

Table 3. Average chemical content of river waters on the continents

Continent	Dry residual (g/L)	Ion (mmol/L)						
		Ca^{2+}	Mg^{2+}	Na^{+}	K^{+}	$[HCO_3]^-$	$[SO_4]^{2-}$	Cl^-
North America	0.142	0.52	0.21	0.39	0.04	1.11	0.21	0.02
South America	0.069	0.18	0.07	0.17	0.05	0.58	0.05	0.14
Europe	0.182	0.77	0.24	0.24	0.04	1.56	0.25	0.20
Asia	0.142	0.46	0.24	0.24	—	1.30	0.09	0.25
Africa	0.121	0.30	0.16	0.48	—	0.69	0.14	0.39
Australia	0.059	0.10	0.12	0.13	0.06	0.52	0.02	0.28
Global average	0.118	0.39	0.17	0.27	0.05	0.96	0.13	0.21

the Amu Daria flows through desert and semidesert regions, the Volga collects its water from mainly nonarid territories.

Table 4 clearly shows that tremendous amounts of soluble salts can be found in the water of some rivers traversing deserts and semideserts. The water-quality problems must be studied individually in such cases in order to monitor, and perhaps change, the chemical composition of irrigation waters.

Table 4 also indicates that in many regions the quantity of airborne salts also should be taken into consideration. However, when those amounts are compared to those collected and transported by waters, it is apparent, with the exception of some seashore districts, that the quantities of airborne salts are negligible. Orlova (1983) states, however, that airborne salts can cause local salinizationin some desert areas.

Among the possible sources of salts, Kovda (1980) also mentions the biological processes, particularly in arid regions where the ash of halophytes may contribute to the salinity of soils and wates. However, it is difficult to say whether it is a cause or a consequence when speaking in terms of landscape geochemistry, because halophytes grow as a result of intensive salinity in their environment.

The main reservoir of salinity in deserts and semideserts is the ground water (Buhoci *et al.*, 1968). Soluble salts accumulated in the ground water as a result of the geochemical processes described above and can cause salinity problems when the ground water is used for irrigation. As long as the groundwater table is deep and moisture cannot rise by capillary flow to the soil profile, even saline ground waters do not cause immediate salinization. As an effect of irrigation, however, the ground water may rise so high that it can reach the surface layers and cause salinization, even when good-quality water is used for irrigation (Darab, 1961). Unfortunately this simple rule has often been ignored during the planning and exploitation of many recent irrigation systems. The hazard of raising the groundwater table by irrigation is underestimated because even when the water table is 10 to 20 m below the surface before construction of the irrigation system, it can easily be elevated to within 1 to 2 m of the surface, particularly where drainage is poor.

The natural or artificial drainage of irrigated land, or land to be irrigated, is also a substantial factor in the processes of secondary

Table 4. Collected salts in the catchment of two rivers per year

River	Catchment area (km^2)	Airborne salts (tonne/km^2)	Discharge to sea (tonne/yr)
Volga	1,401,949	2	8,000,000
Amu Daria	308,804	10	226,000,000

Figure 2. Secondary salinization of irrigation system.

salinization (Maianu, 1964). Drawing an example from the experiences of ancient agriculture, irrigation was practiced for thousands of years in the Nile Valley under a dry climate without secondary salinization. This phenomenon may be explained by the good natural drainage of the land strips along the Nile because of the suction effect of the river. Similar examples have been found recently in many places, e.g., in Syria and Central Asia. As early as the 19th century perennial irrigation was introduced in Egypt. Because of the change in the salt and water regime, the Nile Delta became salinized and an artificial drainage system became necessary to eliminate excess salts (Elgabaly, 1969).

It can be concluded that, with the exception of places with good natural drainage, which are rather rare (Molen and van den Bouhans, 1964), the lack of artificial drainage in deserts and semideserts leads, sooner or later, to secondary salinization. In this contribution the term "secondary salinization" is used in its conventional meaning, as differing from primary salinization which develops without human interference. Secondary salinization develops as a result of human activities, mainly irrigation (Figure 2).

In nonarid areas, where the rainfall and consequently the leaching of substances through soils is remarkable, secondary salinization and alkalization can be avoided, even without artificial drainage (Szabolcs *et al.*, 1969a).

IV. The Development of Irrigation as a Worldwide Method for Increasing Yields and the Consequent Extension of Secondary Salinization and Alkalization

A. Trends of Global Extension of Irrigation

Kovda (1947), Kovda and Szabolcs (1979), Szabolcs (1979), and Thorne and Peterson (1954) have described the main regularities in relation to irrigated soils. The expansion of irrigation not only resulted in more areas being irrigated, but also affected the properties of soils, both positively and negatively. Although irrigation dates back to prehistoric times, its rapid development only started about 200 years ago (Table 5).

From Table 5 it can clearly be seen that the area of irrigated land grew from 8 million ha in 1800 to 48 million ha in 1900 and more than doubled in the last 50 years. This trend is very remarkable and has resulted in large increases in the production of agricultural products but has also resulted in a number of technical and environmental problems. Only a few of these are discussed in this review.

1. At the turn of the century when the irrigated lands of the world did not exceed 50 million ha, they were mainly distributed among a few dry countries, as shown in Table 6. At the beginning of the century the countries listed in Table 6 accounted for two-thirds of all the irrigated soils of the world. At that time irrigation was practiced only in dry regions. However, in recent times, irrigation has been extended into many semihumid regions as well as arid and semiarid regions. The

Table 5. Development of irrigation in the world

Year	Irrigated land ($\times 10^6$ ha)
1800	8
1900	48
1949	92
1959	149
1980	200

Table 6. Irrigated soils in some dry countries in 1900

Country	Irrigated territory ($\times 10^6$ ha)
Indian subcontinent	15.49
Russia	3.80
United States	3.01
Japan	2.72
Egypt	2.00
Italy	1.30
Spain	1.00
Chile	0.30
Total	30.12

problems of salt accumulation and secondary salinization are different among different climatic regions with respect to the chemistry of salt accumulation.

2. In many countries irrigation has been introduced as a new practice within the last few decades. Experiences of countries having long-term irrigation practices were not always known and applied to combat against secondary salinization.
3. The extension of irrigation affected not only the land actually irrigated, but also the neighboring nonirrigated territories. As long as irrigation was concentrated in small areas, evidently its environmental effect was much less than that of big irrigation systems that affected large surrounding areas.
4. The effect of irrigation on the biosphere (besides the irrigated crop) increased parallel with the sharp development of this method, including such adverse effects as:

salinization and contamination of drinking water
water-logged and saline areas as a breeding ground for parasites and diseases
toxic effects on soil microorganisms, etc.

The phenomena listed, and others, constitute in many places a barrier not only to further development of agriculture and human civilization but also to maintaining the present level of production.

B. Effects of Increase of Irrigation on Salt Accumulation in Soils and Waters

The increase of irrigated territories in different parts of the world and in different countries and regions has been remarkable. In the United States, for example, the area under irrigation was doubled between 1949 and 1973 to 21 million ha. Another example: in the Soviet Union every year 1 million ha of new irrigated land are brought under cultivation. In Kenya the area of irrigated land doubled between 1959 and 1969, and further development is envisaged. In Hungary irrigated areas have shown a more than ten-fold increase since the Second World War ended in 1945.

In many countries where irrigation was introduced mainly under nonarid conditions in densely populated areas, its side effects were different from those appearing in most of the arid countries. For example, in dry countries the area surrounding massive irrigated lands is a vast desert, which makes it possible to tolerating such adverse consequences of irrigation as secondary salinization of adjoining areas (Egorov, 1969; Kozlovsky and Kornblyom, 1963). In countries where the utilization of land is over 70%, as in Hungary, the above-mentioned and similar side effects would be catastrophic.

In some countries, such as Egypt, nearly 100% of the agricultural land is irrigated (Elgabaly, 1969). Corresponding figures are; 70% in the Malgache Republic, 26% in Thailand, and 50% in Pakistan (Nazir Ahmad, 1965). Similar ratios exist in many arid and semiarid countries. In less arid or semihumid countries the irrigated land often accounts for only a few percent, but this percentage is sharply increasing even in those countries. For example, in 1980, nearly 13% of the total agricultural land was irrigated in France, more than 10% in Spain, and nearly 15% in Greece (Grande-Covian, 1965, Servant, 1969).

In spite of the availability of many sources of information, accurate data concerning the lands of the world where irrigation has recently been introduced are very scarce. Widely different accounts and estimates can be found in various reports and records. Estimates of total irrigated lands in the world range from 150 to 250 million hectares. The explanation for such diversity of information is probably the fact that it is one thing to register the existing irrigation systems in the world and it is another to keep records of the ones that are in permanent operation. This is the reason, in all probability, that the data of the FAO (Food and Agricultural Organization of the United Nations) and the ICID (International Council for Irrigation and Drainage) are always different regarding the area of irrigated land.

It is evident that neglected or abandoned irrigation systems are rather common and account for a very high percentage of all existing systems. According to the estimates of FAO and UNESCO (United Nations Educational, Scientific and Cultural Organization), as much as half of all the existing irrigation systems of the world are more or less under the influence of secondary salinization, alkalization, and water logging. This phenomenon is very common not only in old irrigation systems but also in areas where irrigation has only recently been introduced.

According to the estimates of all of the above-mentioned agencies, 10

Figure 3. Secondary salinization of irrigated soil.

million hectares of irrigated land are abandoned yearly as a consequence of the adverse effects of irrigation, mainly secondary salinization and alkalization (Figure 3).

The losses and damages are not evenly distributed among the irrigating countries. In some of them the damage is low, but in others it can be so high as to constitute a major problem in the agriculture or even in the national economy of the country. Unfortunately, the world is rich in such sad examples. In Pakistan Nazir Ahmad (1965) carried out statistical analyses respecting secondary salinized land. According to his data, out of 13.8 million ha of total irrigated territory, salinized areas accounted for 2.1 million ha after a few years of irrigation. He indicated among the causes of secondary salinization in Pakistan the joint effect of irrigation and ground water. It is known from FAO reports (1976) and the work of V. Kovda (1980) that more than 50% of irrigated soils in Iraq and Iran are affected by secondary salinization.

According to FAO estimates (FAO, 1971, 1972) for Iraq, 50% of the river plain is affected by salinity. The same authors estimate the extension of salinity and alkalinity in Iran as more than 15% of total land surface, exceding 25 million ha. As a result of irrigation, the salinity problems are increasing year by year.

In Turkey more than 25% of alluvial soils are already affected by salinity and practically all of them may become saline because of the envisaged increase of irrigation (FAO, 1971).

The FAO reports (1971) on salinity in Syria in the Euphrates Valley and estimates the adverse effect as follows:

a. For more than 20,000 ha salinity developed to a level where these soils had to be taken out of cultivation, and the loss is estimated at a total of 30,000 t of cotton (*Gossypium hirsutum*) per year.
b. For about 30,000 ha the yield decreased by 50%, and the total loss is estimated at 20,000 t of cotton per year.
c. For about 60,000 ha the yield decreased by 20%, and the total loss is estimated at about 18,000 t of cotton per year.

The Near and Middle East region depends largely on irrigated agriculture. Although only 36% of the arable land of the regions is irrigated, this area produces 70% of value of all crops. It is envisaged that irrigation will incraese to 51% of the arable area by the next decade. However, the problem of salinity is also expected to increase on these soils because they have a high potential for becoming saline (FAO, 1971).

At present no continent is free from very serious occurrences of salinization and alkalization. According to G. Zavaleta (1965), practically all irrigated alluvial soils in Peru show the features of salinity and alkalinity. In Argentina 50% of the 40,000 ha of land irrigated in the 19th century are now salinized (Kovda, 1980). In Australia secondary salinization and alkalization take place in the valley of the River Murray and in Northern Victoria, where some 80,000 ha have been affected (Jenkin, 1981; Loveday, 1985; Northcote and Skene, 1972). The same phenomenon can be observed in Alberta, Canada (Clayton *et al.*, 1977). Similar processes have been recorded in the northern states of the United States, where irrigation was introduced much later than in the dry west. It must be noted that the last four areas noted and many other irrigated regions are far from being arid, and the majority of salts accumulating are associated with the sodium salts capable of alkaline hydrolysis and not with the neutral sodium salts we are familiar with in desert and semidesert areas (Grin, 1962; Malcolm, 1982; McIntire *et al.*, 1982; Munteanu and Ionescu, 1964).

A great number of sources are available referring to the different problems of secondary salinization and alkalization and describing their adverse effects in many countries. Based on the literature, the following countries can be mentioned where the salinization and/or alkalization of irrigated soils either has represented a serious problem in the past or poses such problems at present. In Europe: Austria (Husz, 1965), Bulgaria, Czechoslovakia, Cyprus, France, Greece, Hungary, Italy, Portugal, (Szabolcs 1979) Rumania (Obrejanu *et al.*, 1964; Sandu *et al.*, 1966), Spain, the Soviet Union (Novikova, 1967; Pekatoros, 1962), and Yugoslavia (Zivkovic, 1965; Miljkovic and Plamenac, 1971; Szabolcs, 1979); in North America: Canada and the United States (Richards, 1954); in Mexico and Central America: Cuba and Mexico; in South America: Argentina, Brazil, Chile, Columbia, Peru, and Venezuela; in Africa: Algeria, Angola, Chad, the Cameroons, Egypt, Ethiopia, Ghana, Kenya, Libya, Morocco, Niger, Nigeria, Somalia, South West Africa, Sudan, Tanzania, Tunisia, Zambia, and Zimbabwe; in the Near and Middle East and southern Asia: Afghanistan, Burma, India, Iran, Iraq, Israel, Jordan, Kuwait, Lebanon, Pakistan, Saudi Arabia, Syria, the Trucial States, Turkey and Yemen; in North and Central Asia: China, Mongolia, and the Soviet Union; in Southeast Asia: Indonesia, Malaysia, Thailand, and Vietnam; and in Australasia: Australia (Peck *et al.*, 1983; Pels and Stannard, 1977). According to the knowledge of and the data available to the author, in these countries secondary salinization not only occurs but represents a practical problem. The absence from the list of several countries where the phenomenon may also occur is because of missing records or the lack of information of the author of this paper (Szabolcs, 1979).

On the basis of the available data, we can conclude that parallel with the accelerating increase of irrigation in many parts of the world the processes of secondary salinization and alkalization also accelerate. We have—even if not fully exact—reliable data on the extent of existing salt-affected soils on our globe, which is shown in Table 7.

The total figure indicates that nearly one tenth of the land area of our

Table 7. Salt-affected soils on continents and subcontinents

	$\times 10^6$ ha
North America	15.755
Mexico and Central America	1.965
South America	129.163
Africa	80.608
Southern Asia	87.608
North and Central Asia	211.686
Southeast Asia	19.983
Australasia	357.330
Europe	50.804
Total	954.832

globe is touched with soil salinity and/or alkalinity. A certain part of this territory is under irrigation but the greater part or irrigated areas consists of soils that are nonsaline and nonalkali, at least for the moment. Part of such land may be potentially saline or potentially alkaline. Potential salinity or alkalinity means that a soil that is neither saline nor alkaline at the moment can be turned into either by applying improper methods of irrigation. Most of the present secondary salt-affected soils have gone through this process (Szabolcs, 1979).

While we have reliable records on the extension of salt-affected soils in our globe, unfortunately we still lack proper data on the world extent of secondary salinized or alkalized soils; we must make our own estimates. The above-cited UNESCO/FAO data on 50% salinization and alkalization of irrigated areas provide some guidance for making such calculations. Based on these and other figures, we can conclude that the territories of secondary salinized and alkalized soils that once were fertile but have since been turned into bare deserts, as a side effect of improper irrigation, are on the same order of magnitude as those recently existing salt affected soils that have never been irrigated. Such estimates are essential to any extrapolation into the future.

Neither do we have proper data and records on the extent of potentially salt-affected soils. It is necessary first of all to clarify this term. As mentioned above, potential salt-affected soils are those which are nonsaline or alkaline on the top layers at the moment but may be salinized by irrigation. Obviously, such a definition is relative because any soil can be salinized, e.g., when irrigated with saline water or in lack of any drainage (Talsma, 1963).

Kovda *et al* (1967) stressed that in different regions of the world widely varying soil and subsoil conditions exist. The relationships between irrigation, crops, and soil must still be developed fully under such conditions (Poljakoff *et al.*, 1975). In soils to be irrigated a salt and water balance exists between soil, crop, and irrigation and is influenced by the rainfall, stream flow, ground waters, and evaporation and transpiration. This balance is altered during irrigation.

According to the cited and other authors (Anayan, 1971; Borovsky, 1982; Szabolcs *et al.*, 1966; Boumans and van der Molen, 1964; Zavaleta, 1965) ground water plays a significant role in salinity because of seepage from channels during irrigation. Large amounts of irrigation and drainage water seep into ground water from unlined canals, lifting the water table near to the surface of irrigated and adjoining soils. There are situations where this contribution to ground waters may be as large as and sometimes even larger than the quantity actually utilized by irrigated crops.

This new accretion to ground waters may be withdrawn from the subsoil, wholly or in part, by wells or tube wells. Under good subsoil drainage conditions, it may seep into rivers. Under other conditions, it may raise the subsoil water level or fill pockets or form perched water tables and cause local waterlogging, necessitating drainage for restoring healthy conditions in the soil.

Consequently irrigation also changes the salt content and salt distribution in the soil profile (Szabolcs, 1979; Szabolcs and Darab, 1968). That is why the definitions of secondary salinization and alkalization should always be examined against the background of the methods of irrigation, soil and water properties, farming pattern, natural or artificial drainage, etc. This is the reason that diagnosis of secondary salt-affected soil based on a simple soil survey is always difficult and often omitted, even during the planning stage of irrigation. This omission has caused great unexpected harm in the first years or later periods of exploitation of irrigation systems in many countries. Evidently, under different climatic conditions secondary salinization has different interpretations. Closely related to the biogeochemical processes of salt accumulation, the hazard of secondary salinization evidently is greater in desert areas than in humid regions where natural leaching processes remove soluble salts.

Survey and estimation of possible potential salt-affected soils were carried out in some European countries (Szabolcs, 1974). The figures are shown in Table 8.

The data in Table 8 clearly demonstrate that even in those European countries where the hazard of salt accumulation is far less than for some countries discussed earlier, the area of potential salt-affected soils is similar to, or greater than, the area of recent salinization. Under arid conditions, obviously, this ratio will be even higher.

In many arid and semiarid areas practically all soils, or at least a high percentage, can be termed potentially saline. As a consequence of the regularities noted above, the determination, grouping, characterization, and mapping of secondary salt-affected soils must be performed in the context of the local environmental and economic conditions (Ianovici and Florea, 1964).

Table 8. Existing and potential salt-affected soil in some European countries

Country	Salt-affected soils (ha)	
	Existing	Potential
Austria	500	2,500
Czechoslovakia	25,000	80,000
Hungary	740,000	885,000
Italy	500	400,000
Soviet Union	28,000,000	18,000,000

Apart from environmental conditions, econcomic aspects must also be taken into consideration. This is because even though the prevention of secondary salinization is theoretically possible in many developing countries, in most cases the construction of expensive drainage systems can make doubtful the economic expediency of the introduction of irrigation (Kizilova, 1961; Kolpakov, 1961).

Although irrigation is the principle cause of secondary salinization on a global scale, it is worthwhile to list other causes that have been studied and documented (Kovda, 1980).

1. When nonirrigated land is utilized
   a. the formation of secondary alkaline or saline soils resulting from overgrazing and compactness of sod meadow land:

      along the contact belt of mountainous foothills and plains,
      on low terraces of valleys after flooding by rivers was excluded by barrier construction

   b. the formation of secondary saline soils as a result of disposal of brackish water pumped from:

      petroleum wells
      coal mines
      industrial plants

   c. salinization of soils after seawater invasion under the influence of land subsidence, or after heavy tsunami or storms and earthquakes
   d. accelerated formation of saline alluvial soils on deltaic and tidal wave territories after periodic floodings have stopped as a result of dam construction in the middle and upper reaches of the river valley
2. When irrigated land is utilized:
   a. the formation of water-logged and saline soils along unlined canals as a result of water seepage, groundwater elevation, and evaporation
   b. the formation of spotty saline fields, then totally saline fields after 5 to 15 years of irrigation without appropriate drainage installations for the evacuation of saline subsoil water
   c. wrong application of methods for irrigating soils with brackish (or alkaline) irrigation water taken from:

      saline rivers
      tube wells installed into saline ground water or after overpumping good-quality subsoil water
      sea or gulf sources

   d. appearance of saline soils on valley terraces above and after the construction of a dam (as a result of the submerging of subsoil water following reservoir formation).

Based on this, we must agree that manmade salinization is one means to the destruction of the global biospheric mechanism, with an influence not only directly on the soil but also indirectly on several processes ranging from photosynthesis to the cycling of bioelements (C, O, N), etc. Such influences must also be taken into account with respect to soil organic matter, energy resources, and soil bioprocesses, etc. The negative consequences of soil salinization are not only social and economic; they are globally destructive for the biosphere of our planet (Kovda, 1980).

Some recent studies describe the adverse effects of salinity not only on plants but on animals, too. The increased electrolyte content of the environment, particularly of waters and soils, results in excessive salt intake by animals, disturbing the metabolism of their life functions. In Australia, for instance, sheep and other livestock die from time to time after drinking salty water in hot regions. Kidney and other organs suffer from saline and sodic waters; the total body water increases and there is an expansion of extracellular volume. Other disorders also occur in the wake of the salinization of soil and water, particularly in dry areas (Peck *et al.*, 1983).

V. Combating Secondary Salinization and Alkalization

The extension of irrigation remains a major prospect for increasing yields and for nourishing the world's population. If care is taken not only in the exploitation of irrigation systems but also during the planning and construction phases of irrigation systems, if the necessary preliminary surveys and precautions are carried out and taken into consideration, respectively, then the production of food and raw materials can be multiplied in the future on a worldwide basis.

Many prognoses are available concerning the development of irrigation for the turn of the millenium and for the 21st century. Some of them are local or national reports, but some of them are on a global scale, such as the well-known Report for the President and ICID prognoses (Alekseevsky, 1974). It is evident that in different sources different figures can be found, but on the average about 400 million ha of irrigated land are predicted for the first part of the 21st century. Unfortunately no reliable predictions are available on the hazard of developing secondary salinization resulting from such a sharp increase in irrigated land in the near future. Based on experience, it must be agreed with many authors that in general the hazard of secondary salinization and alkalization does not increase in linear proportion with the increase of the acreage of irrigated land: the correlation is closer to logarithmic. We still lack exact analysis of the rate of the possible hazard of recent and predicted extension of irrigation in different countries (Szabolcs and Darab, 1982).

Effects obviously must be very diverse in different areas, regions, and districts, but if secondary salinization increases in space and extent as a worldwide process, its global importance is more than likely to increase sharply in the future.

It is relevant to develop, whenever possible, and to adopt different studies and their results in the local, national, and even international planning of new irrigation systems as soon as possible. It is also necessary to extend such studies to predicting the joint effect of existing and future irrigation systems on the environment.

In case of future development of irrigation, not only the pedological but also the general environmental influence of irrigation will assume new dimensions. For instance, the concentration of CO_2 in the atmosphere, and its rapid change as effected by human activities, is a widely discussed problem of our days. A great number of books, reports, and prognoses are available on this subject; some of them threaten consequences that would be tragic for mankind. If the area of irrigated land doubles or trebles, irrigated plants will be able to produce higher amounts of biomass and harvest, through photosynthesis, and to consume as much as 30 to 40 billion tons of CO_2 annually, instead of the recent 15 to 20 billion tons (Kovda, 1980).

The technical literature presents many examples clearly showing that the successes of irrigation are also interrelated with numerous vital problems. As is clear from the examples mentioned, the aim of irrigation development is to improve the food situation as well as the environment. It is also evident that the hazard of secondary salinization and alkalization will be one of the major obstacles in the way of this development if we do not intensify study of this risk and apply methods for its prediction and prevention.

It is a must to intensify both the theoretical and technical activities relating to the hazard of secondary salinization and alkalization. Until now the influence of the geochemical and hydrogeochemical processes have been underestimated in many places, which sooner or later results in salt accumulation by irrigation in the given territory. Although an abundance of studies and quality requirements for irrigation water is available (Szabolcs and Darab, 1982), the effect of ground water on soil salinity is often left out of consideration. That is why a comprehensive preliminary study is necessary before planning new and extending existing irrigation systems, not only of soils and surface waters, but also of underground waters and layers (Darab and Ferencz, 1969).

In relation to the influence of ground water on salt accumulation, first of all the so-called critical depth of the groundwater table should be determined. This level means the depth below which, owing to natural or irrigated conditions, leaching prevails, whereas above this level salt accumulation takes place in the soil profile. In other words, the salt regime of the given territory is in equilibrium at the critical depth.

A. Salt Regime

The salt regime of soils denotes the accumulation and the leaching of the soluble salts, i.e., the periodic changes of these processes in the soil. Examination of the salt regime, the periodical changes in the quantity and quality of salts, reveals the prevailing direction of their movements and whether accumulation or leaching of salts takes place under the given conditions.

B. Salt Balance

Although study of the salt regime supplies important data on the dynamics of a salt-affected soil, it does not present in itself any further information. Therefore the next step is to compare the salts contents of the soil measured at given times and to express them in salt balances. Salt balances reveal the effects, or the effects to be expected, of amelioration of agrotechnical measures on the changes in the degree of salinity and alkalinity of the soil as well as those of irrigation on its soluble salt contents. The extablishment of salt balances requires an exact knowledge of the cause, speed, and degree of accumulation or leaching.

The following data are needed to establish the salt balance, regardless of the extent of the area concerned:

1. total amount of soluble salts at the beginning and the end of the observation
2. the increase of soluble salt contents during the observation
3. the decrease of soluble salt contents during the observation

Three types of salt balances may be distinguished:

1. stable salt balance
2. balance of salt accumulation
3. balance of leaching

The salt balance of soils may be established for both larger (e.g., a drainage basin) and smaller territorial units.

A salt balance may be established either for the whole soil profile or for given genetic horizons.

The salt balance of a soil depends on the joint effect of numerous factors. Among these factors the following must be specifically mentioned:

C. The Depth and Chemical Composition of the Ground Water

When considering the water table, we must take into account the fact that it fluctuates periodically. The depth of the water table is influenced by local factors, such as the amount of precipitation in the given place and the micro- and mesorelief of the area. It is also affected by the amount of water getting into the catchment area from various sources that contributes to the ground water, and by the time that it takes for the water to manifest itself in the place examined. The depth of the ground water also changes when the natural conditions are artifically altered by the effects of—e.g., inland—drainage or irrigation.

Analysis of the chemical composition of the ground water must be carried out simultaneously with the determination of the water table depth.

The following qualitative data on the ground water must be given:

1. the total salt contents of the ground water (in mg/L)
2. the relative amount of sodium salts in the ground water
3. the relative proportion of magnesium salts within the alkali earth metal salts
4. the distribution of soluble salts in the ground water as measured by the anions

The physical and water regime properties of a soil also affect its salt regime to a considerable extent. Of these, two should be mentioned specifically: the water conductivity of the soil profile and, if present, the depth of the accumulation horizon of heavier mechanical composition or of disadvantageous water regime properties.

D. The Influence of Relief

Factors that should be determined are:

1. the infiltration of rainfall or irrigation water into the soils of different relative depths
2. the wetting of the soil profile
3. the moisture content of deeper layers when the downward movement of rainfall or irrigation water is followed by the upward movement of ground water

E. Irrigation Techniques

Under irrigated conditions the salt regime of the soils also depends on the method and the technical level of irrigation management.

In the case of surface irrigation, as a function of the amount of the applied irrigation water, excessive leaching may result in raising the water table, decreasing the possibilities for leaching soluble salts, and may even increase the salt balance of the soils. This is particularly true when the drainage conditions of the irrigated area are unsatisfactory.

F. Calculation of Salt Balance

Given the knowledge of the factors influencing the increase and decrease of the salt contents of soils, the salt balance may be established on the basis of the following equation:

$$b = a + \left(d + \frac{cv}{m\ t_{fs}} \cdot 10^{-5} \right)$$

Where:

b = soluble salt content of the soil at the end of the observaiton (in mg/100 g soil)
a = soluble salt content of the soil at the beginning of the observation (in mg/100 g of soil)
c = the salt concentration of the irrigation water (in g/L)
v = the quantity of the irrigation water applied during the observation period (in m^3/ha)
M = the thickness of the soil layer for which the salt balance was established (in m)
t_{fs} = the bulk density of the soil
d = the salt regime coefficient of the soil (in g/100 g of soil)

The change that has occurred in the salt content of a soil during the observation period is expressed in the salt regime coefficient. This coefficient gives the difference between the amount of salts leached from the soil and that of salts which got into the soil from sources other than the irrigation water.

As it is rather difficult to calculate the value of d, we had better determine the change in the salt content of the soil and calculate the salt regime coefficient on the basis of the following correlation:

$$d = b - \left(a + \frac{cv}{M\ t_{fs}} \cdot 10^{-5} \right)$$

Having completed the calculations for several irrigated soils, we can distinguish the following characteristic cases:

1. The salt regime constant is a negative value.
2. The salt regime constant is a negative value but the salt content of the soil increased during the observation period. This means that although leaching took place to a certain extent, more salt was introduced with the irrigation water than could be leached out.
3. The salt regime constant is a positive value and the soluble salt content of the soil increased during the observation period. This indicates that more salt got into the soil—not only from the irrigation water but also from the ground water and perhaps from other sources, too—than could be leached out.

4. The salt regime coefficient of the soil may be zero. In this case the salt balance of the soil is stable and the current depth of the groundwater table is practically tantamount to the critical depth.

To calculate the salt regime coefficient with the above simple equation requires the regular determination of salt movements in the given territory.

Where even such determinations are difficult, Kovda (1980) recommends a simple empirical estimate of the critical depth of the ground water:

$$L = 170 + 8t \pm 15$$

Where:
L = the critical depth (in cm)
t = the mean annual temperature (degrees centigrade)

G. Salinity Hazard

For controlling the possible hazard of salinization and/or alkalization in irrigated areas or areas to be irrigated, the following factors should be studied and detemined:

1. climatic factors, such as temperature, rainfall, humidity, vapor pressure, evaporation, and their fluctuations and dynamics
2. geological, geomorphological, geochemical, hydrological, hydrogeological, and hydrochemical factors, such as natural drainage, the depth and fluctuation of the water table, and the direction and velocity of horizontal groundwater flow, and the salt content and composition of the ground water
3. soil factors, such as soil profile, texture, structure, saturated and unsaturated water conductivity, soluble salt content, salt composition and salt profiles, exchangeable cations, and pH
4. agrotechnical factors, such as land use, crops, and cultivation methods
5. irrigation practices, such as the amount of irrigation water; the method, frequency and intensity of irrigation; salt content and composition of irrigation water; and natural and artificial drainage

H. Preliminary Survey and Control of Irrigated Soils

The above-mentioned factors determine the aims and methods of the preliminary survey of soils made in order to define the degree or the existence of potential salinity and/or alkalinity.

The soil properties, groundwater depth, and chemical composition, as well as the salt balance and hazard, can be demonstrated on maps with recommendations about the technique of irrigation and water use.

Obviously, environmental conditions, on the one hand, and the methods of utilizing the territory in question, on the other hand, should be taken into consideration when an area is evaluated in this respect. Because of this, different limit values and different methods—based on uniform principles—should be selected in the course of this procedure.

In Table 9 a scheme of methods recommended for the control of salinity and alkalinity in irrigated areas is given.

This table shows that the prediction of secondary salinization and alkalinization of the soils to be irrigated should be based on a preliminary survey of the landscape and soils before the construction of the irrigation system. In this way, it is possible to take the necessary steps for the prevention of adverse processes.

During irrigation, a well-organized monitoring of the soil and water properties is to be conducted in order to record changes, if any, and to underlie taking precautions, if necessary. Monitoring methods, as well as the timing and location of sampling, depend on local conditions.

In the course of making the survey and monitoring, in order to develop a reliable method for the prediction of salinization and alkalinization, the following problems must be solved:

1. The main sources of water-soluble salts (irrigation water, ground water, surface waters, salty deep soil layers, etc.) must be identified.
2. The main features of the salt regime must be characterized (salt balance); and the whole range of natural factors influencing the salt regime must be analyzed.

Consequently, an exact salinity and/or alkalinity prognosis must be based on the evaluation of many natural and human factors and a through knowledge of the existing soil processes.

Table 9. Scheme of methods recommended for the control of salinity and alkalinity in irrigated areas

A. Preliminary survey before construction of irrigation system	
Landscape	*Planned irrigation*
Climate	available irrigation water, quality and quantity
Hydrology	groundwater depth and quality
Hydrogeology	technology of irrigation
Geomorphology	cropping pattern tolerance
B. Monitoring during irrigation	
Salinity and alkalinity of soil and groundwater table	
Chemical composition of ground water	
Chemical composition of irrigation water	
Physical soil properties	
Toxic elements, if any, in soil and water	

Numerous handbooks and recommendations are available on quality requirements for both soils and waters as the subject and means of irrigation, respectively. We must note that any of them must be adjusted to local conditions (Szabolcs and Darab, 1982). It is beyond the aims of this contribution to compare and evaluate the different irrigation water-quality systems of different countries and regions; as a general rule, however, the following properties can be listed as the subjects of determination for all systems.

1. total salt contents of irrigation water
2. sodicity (Na^+) of irrigation water
3. alkalinity of irrigation water
4. Magnesium content of irrigation water
5. Boron content of irrigation water

Mapping of the results of preliminary and subsequent surveys constitutes not only a good display of soil and environmental conditions of the irrigated areas, or areas to be irrigated, but also guideline for proper irrigation and land protection. Such systems elaborated by various authors, for the different places and conditions are also available in the technical literature (Szabolcs *et al.*, 1969a, b).

The monitoring system of irrigated areas must be elaborated on and/or adapted to, and also closely related to, local circumstances. Soils, irrigation water, and ground water must be studied regularly, and whenever discrepancies occur with the predicted salt regime the necessary measures should be taken either by diminishing the acreage or the intensity of irrigation or by improving the drainage of the land.

In this short review only a few examples are furnished of the methods of studies developed to predict and prevent secondary salinization and alkalinization of irrigated soils.

VI. Conclusions

The accelerated growth of the extent of irrigated soils foreshadows the increasing hazard of secondary salinization and alkalinization. This adverse process which kept track with the whole development of irrigated agriculture, deteriorated during thousands of years on many millions of hectares of once fertile land. Recently it poses a problem of global importance, threatening our fertile soils, our waters, and plant, animal and human health.

Irrigation is extending not only in arid regions, but also in semihumid and humid regions, where the environmental conditions demand irrigation technologies different from those practiced in desert conditions. Accordingly the combat of secondary salinization and alkalization should be specific for climatic as well as technological conditions.

In spite of general acceptance and awareness of the global hazard of secondary salinization and alkalization, these processes have not been arrested or diminished. On the contrary, they are expanding. This fact is reflected in a great number of books, papers, and reports devoted to the subject.

The weakness is not always the lack of sufficient study, but rather the lack of sufficient knowledge of how to adopt methods that are both technically sound and economically feasible. One of the main tasks for the future is to intensify the study of the possible hazard of increasing irrigation on soils and the environment and to elaborate up-to-date and economical methods for combatting the growing danger of the development of harmful processes.

References

Alkeseevsky, E.E. 1971. *Irrigation and drainage of the world*. Izd. Kolos, Moscow. In Russian.

Anayan, A.K. 1971. *Drainage for the amelioration of soda solonchaks* (USSR). Izd. Kolos, Moskva. In Russian.

Arany, S. 1956. *Salt affected soils and their reclamation*. Mezögazdasági Kiadó, Budapest.

Aubert, G. 1962. Arid zone soils. A study of their formation characteristics, utilization and conservation. *The problems of the arid zone, Proc. Paris Symp. Arid Zone Res.* 18:115–138.

Bazilevich, N.I. 1965. *Geochemistry of sodic soils*. Izd. Nauka Moskva. In Russian.

Borovsky, V.M. 1982. *Formation of salt affected soils and the hydrogeology in Kazakstan. Alma Ata*, Publishing House Nauka of Kazakh SSR. In Russian.

Borovsky, V.M., and M.A. Pogrebinsky. 1964. Soil formation and soil amelioration in continental deltas (USSR) In Russian. *Plod. Melior. Pochv. SSSR. (Dokl. Mezhd. Kongr. Pochv.)* 8:173–181.

Boumans, J.H. and W.H. Molen van der. 1964. Drainage requirements of irrigated soils in relation to the salinity. *Landb. Tijdschr.*, 76:880–887.

Buchoci, L., C. Ionescu, and T. Covaliov. 1968. Soil and groundwater salinization in the dammed enclosure from Calmatui. *Gropeni Studii Hidroamelior.* 4:263–281.

Buringh, P. 1960. *Soils and soil conditions in Iraq*. Baghdad.

Clayton, J.S., W.A. Ehrlich, D.B. Cann, J.H. Day, and I.B. Marshall. 1977. *Soils of Canada*, Vols. I and II. Canada Dept. of Agriculture, *Ottawa*.

Darab, K. 1961. Salt balance and salt regime of irrigated soil in Hungary. *Agrokémia és Talajtan* 10:305–314.

Darab, K. and Ferencz, K. 1969. *Soil mapping and control of irrigated areas* (Hungary). OMMI Publ. Ser. 1, No. 10, Budapest.

Dregne, H.E. 1976. *Soils of arid regions*. Elsevier, Amsterdam.

Egorov, V.V. 1969. Some aspects relating to the sodic salinization in the subhumid regions of Europe and Asia. In Russian. *Agrokémia és Talajtan* 18:Suppl. 187–191

Elgabali, M.M. 1969. Three types of sodic soils in the United Arab Repbulic. *Agrfokémia és Talajtan* 18:Suppl. 87–99.
FAO 1971. *Irrigation and Drainage* Paper 7; Rome.
FAO 1972. *Irrigation and Drainage* Paper 13; Rome.
Fersman, A.E. 1934. *Geochemistry*. Leningrad.
Grande-Covian, R. 1965. Reclaiming the Guadalquivir marshes. *Fatis Rev.* 12(3):73–76.
Grin, G.S. 1962. Salt affected soils in the Ukraine and their origin. *Trudy Kharkov sel'khoz. Inst.* 39:8–102. In Russian.
Hodge, C., and P.C. Duisberg. 1963 *Aridity and man*. American Association for the Advancement of Science, Washington, D.C.
Husz, G. 1965. Theory and practice of the amelioration of salt affected soils with special regard to conditions in the "Seewinkel" region (Austria) I. Theory. *Bodenkultur* 16:223–244.
Hyams, E. 1976. *Soils and civilization*. John Murray, London.
Ianovici, V., and N. Florea. 1964. The accumulation of salts in the soils of quaternary plains in Rumania. *Studii Pedol.* C 14:5–20.
Jenkin, J.J. 1981. Terrain, groundwater and secondary salinity in Victoria, Australia. *Agric. Water Manag.* 4:143–171.
Kizilova, A.A. 1961. Movement of easily soluble salt in solonchak soils under leaching. *Salinity problems in the arid zones. Proc. Teheran Symp. Arid Zone Res.* 14:227–232.
Kolpakov, V.V. 1961. Effect of a water table rise on evaporation and the water soluble salt regime of soils in the Presivash region (USSR). *Izv. Timiryazev sel'khoz. Akad.* 1:132–147. In Russian.
Kovda, V.A. 1947. *Origin and regime of salt affected soils*, Vols. I and II. Izd. Akad. Nauk. SSSR, Moscow. In Russian.
Kovda, V.A. 1980. *Problem of combatting salinization of irrigated soils*. UNEP. NAIROB1.
Kovda, V.A., and N.G. Minashina (Eds.). 1967. *Irrigation and drainage of salt affected soils* (USSR). Izd. Nauka, Moskva. In Russian.
Kovda, V.A., and I. Szabolcs. 1979. Modeling of soil salinization and alkalization. *Agrokémia és Talajtan* 28:Suppl. 351–380.
Kovda, V.A., C. van den Berg, and R.M. Hagan. 1967. *International source book. Irrigation and drainage of arid lands. Salinity and alkalinity*. FAO-UNESCO, Paris.
Kozlovsky, F.I., and E.A. Kornblyum. 1963. Soil ameliorative conditions in the Volga-Akhtuba bottomland (USSR) in connection with its development and evolution. *Pochvovedenie* 7:73–84. In Russian.
Loveday, J. 1985. . Soil salinity and sodicity in Australia. *Agrokémia és Talajtan* 1–2. (In press).
Maianu, A. 1964. *Secondary soil salinization* (Rumania). Edition of Academy of Rumanian Socialist Republic. Bucharest.
Malcolm, C.V. 1982. Wheatbelt salinity. A review of the salt land problem in South Western Australia. Bull. No. 52, Western Australian Department of Agricultural Technology, 65 pp.
McIntyre, D.S., J. Loveday, and C.L. Watson. 1982. Field studies of water and salt

movement in an irrigated swelling clay soil. III. Salt movement during ponding. *Austr. J. Soil Res.* 20:101–105.
Miljkovic, N. and N. Plamenac. 1971. Solonetz soils of Yugoslavia, their properties and possibilities of utilization. In: I. Szabolcs (ed.), *European solonetz soils and their reclamation*. Akadémiai Kiadó, Budapest, pp. 151–164.
Molen, W.H. and J.H. van der Boumans. 1964. Drainage requirements of irrigated soils in relation to salinity. *Trans. 8th Int. Cong. Soil Sci.* 2:847–854.
Munteanu, I., and M. Ionescu. 1964. Bog soils with maritime salinization in the Danube delta (Rumania). *Trans. 8th Int. Cong. Soil Sci.* 5:667–674.
Nazir Ahmad. 1965. A Review of salinity-Alkalinity status of irrigated soils of West Pakistan. *Agrokémia és Talajtan* 14:Suppl 117–154.
Novikova, A.V. 1967. Prognosis of secondary salinization and alkalization of soils due to irrigation (USSR). *Agrokhim. Poch.* 5:3–9.
Northcote, K.H., and J.K.M. Skene. 1972. Australian soils with saline and sodic properties. Soil Publ. No. 27, CSIRO Australian Division of Soils, Adelaide.
Obrejanu, G., A. Maianu, and I. Aksenova. 1964. Salt accumulation in mineralized groundwaters and saline soils of floodplains in the lower Danube bottomland (Rumania). *Pochvovedenie* 8:44–62.
Orlova, M.A. 1983. Aeolic factors in salt accumulation. Alma-Ata, Publishing House Nauka of Kazakh SSR. In Russian.
Peck, A.J., J.F. Thomas, and D.R. Williamson. 1983. *Salinity issues.* A.G.P.S. Canberra.
Pekatoros, L.G. 1962. Secondary salinization of soils in the Dneper left bank bottomlands and Danube delta, in the Ukraine. *Pochvovedenie* 2:26–36. In Russian.
Pels, S., and M.E. Stannard. 1977. Environmental changes due to irrigation development in semi-arid parts of New South Wales, Australia. In: E. Barton Worthington, (ed.), *Arid land irrigaiton in developing countries, environmental problems and effects*. Pergamon Press, pp. 171–183.
Poljakoff, Mayber, and Gale, J. 1975. *Plants in saline environments.* Springer Verlag, Berlin-Heidelberg.
Polynov, B.B. 1956. *Selected papers.* Izd. Akad. Nauk SSSR, Moscow. In Russian.
Prognosis of salinity and alkalinity. Soil Bulletin 31, FAO, Rome.
Richards, L.A. 1954. Diagnosis and improvement of saline and alkali soils. Handbook No. 60, U.S. Department of Agriculture, Washington, D.C.
Rozanov, A.N. 1961. System of scientific investigations in projects for the irrigation and reclamation of saline soils. In: *Salinity problems in the arid zones, Proc. Teheran Symp. Arid Zone Res.*, 14:223–226.
Sandu, G. 1966. Study of ground water mineralization and the hydrosaline regime of soils in the Calarasi terrace irrigation systems (Rumania) An. Sect. Pedol. 1965. 33:323–343.
Servant, J. 1969. *Main characteristics of saline and alkali soils in the Roussillon Plain Bull.* Assoc. Fr. Etude du Sol. No. 3, Paris.
Shainberg, I., and J. Shalhevet. 1984. Soil salinity under irrigation. Ecological studies. Analysis and synthesis, Vol. 51. Springer Verlag, Berlin.
Szabolcs, I. 1974. Salt affected soils in Europe. Martinus Nijhoff, The Hague, and

Research Institute for Soil Science and Agricultural Chemistry of the Hungarian Acadamy of Sciences, Budapest.

Szabolcs, I. 1979. *Review on research of salt affected soils*. UNESCO, Paris.

Szabolcs, I., and K. Darab. 1955. Dynamics of soluble salts in irrigated soils (Hungary). *Agrokémia és Talajtan* 4:251-264.

Szabolcs, I., and K. Darab. 1968. Salt balance and salt transport processes in irrigated soils (Hungary). *9th Int. Cong. Soil Sci.* 1:491-498.

Szabolcs, I., K. Darab, and G.Várrallyay. 1966. Salt balance of irrigated soils (Hungary). *Beitr. Trop. Landw. Vet. Med.* 4(2):123-135.

Szabolcs, I., and K. Darab. 1982. Irrigation water quality and problems of soil salinity. *Acta Agron. Sci. Hung.* 31(1-2):173-194.

Szabolcs, I., K. Darab, and G. Várallyay. 1969a. Salt balances for the prediction and prevention of secondary salinization and alkalinization of irrigated soils (Hungary). *Nemzetk. Mezögazd. Szemle* 13(5):46-50.

Szabolcs, I., K. Darab, and G. Várallyay. 1969b. Methods for the prognosis of salinization and alkalinization due to irrigation in the Hungarian Plain. *Agrokémia és Talajtan* 18:Suppl. 351-376.

Talsma, T. 1963. Control of saline ground-water. *Meded. Landbouwhogeschool, Wageningen*, 63(10):1-68.

Thorne, D.W., and H.B. Peterson. 1954. *Irrigated soils*. Blakston Co., New York, Toronto.

UNESCO. *Salinity problems int he arid zones. Proc. Teheran Symp. Arid Zone Res.*, 14:1-395. (UNESCO, Paris, 1961.)

USSR/UNEP Project, 1982. *Studies on the impact of agricultural management on the environment. Summary of research results*. Moscow.

Worthington, E.B. 1977. *Arid land irrigation*. Pergamon Press, ICSU.

Zavaleta, G.G. 1965. The nature of saline and alkaline soils of the Peruvian coastal zone. *Agrokémia és Talajtan* 14:Suppl. 415-425.

Zivkovic, B. 1965. Salinization and comparative characteristics of normal soils, soils in the process of salinization and solonchaks in Vojvodina (Yugoslavia). *Savremena poljoprivreda* 5(Spec. ed.):1-91.

Index

N

O

P

MIX
Papier aus verantwortungsvollen Quellen
Paper from responsible sources
FSC® C105338

If you have any concerns about our products,
you can contact us on
ProductSafety@springernature.com

In case Publisher is established outside the EU,
the EU authorized representative is:
Springer Nature Customer Service Center GmbH
Europaplatz 3, 69115 Heidelberg, Germany

Printed by Libri Plureos GmbH
in Hamburg, Germany